Third Edition

Practicing Biology

A Student Workbook

Campbell • Reece

Biology

Eighth Edition

Jean Heitz and Cynthia Giffen

University of Wisconsin, Madison

PEARSON

Benjamin
Cummings

San Francisco • Boston • New York
Capetown • Hong Kong • London • Madrid • Mexico City
Montreal • Munich • Paris • Singapore • Sydney • Tokyo • Toronto

Editor-in-Chief: Beth Wilbur
Senior Editorial Manager: Ginnie Simione Jutson
Senior Supplements Project Editor: Susan Berge
Project Editor: Strawberry Field Publishing, Melanie Field
Executive Marketing Manager: Lauren Harp
Managing Editor: Michael Early
Production Supervisor: Jane Brundage
Manufacturing Buyer: Michael Early, Michael Penne
Production Service: S4 Carlisle Publishing Services, Erin Melloy
Cover Design: Yvo Riezebos Design
Text and Cover Printer: Courier, Stoughton
Cover Photo Credit: Magnolia Flower–Corbis. Photographer: Chris Fox

ISBN-13: 978-0-321-52293-1
ISBN-10: 0-321-52293-1

Contents

About Practicing Biology: A Student Workbook

Contents

About Practicing Biology: A Student Workbook

What does *Practicing Biology: A Student Workbook* contain?

The activities in this workbook focus on key ideas, principles, and concepts that are basic to understanding biology. The overall organization follows that of *Biology*, 8th edition. Key principles or processes developed in the activities are often revisited and integrated into subsequent activities. While the individual activities may vary in the thought processes required and in their specific biological content, the overall goals of this workbook are to:

- Allow you to discover what you know and, more important, what you don't know.
- Help you to discover and modify any misconceptions that may exist in your understanding of biology.
- Provide you with opportunities to synthesize and apply what you learn to novel situations.

What kinds of activities are included?

The activities in *Practicing Biology* take a number of different forms.

Leading Question. In these activities, you are asked a series of leading questions that are designed to build your basic understanding of principles. Leading questions are generally followed by additional questions that give you the opportunity to apply what you have learned to new situations.

Concept Mapping/Diagramming and **Drawing.** These activities, which include drawing exercises and the development of concept maps and flow diagrams, are designed to help you organize information and ideas and develop an understanding of how various pieces of information are interrelated.

Modeling. Modeling activities provide you with instructions for building models of dynamic biological processes that occur at the molecular, cellular, and physiological levels. Modeling can help you to both develop and test your understanding of processes that are generally invisible to the naked eye.

Process of Science. Process of science activities allow you to practice the use of scientific thought processes and are designed to give you a better understanding of how the knowledge you gain in class can be applied to:

- propose experiments,
- predict possible outcomes of experiments, and/or
- interpret experimental data.

Reviewing. This group of activities provides an opportunity for you to review and integrate key ideas and principles in biology. The reviews are generally followed by activities that require you to apply your knowledge.

Teaching. In the teaching activities, you will have a chance to examine ideas, principles, and concepts from the instructor's point of view. Most instructors will agree that your understanding of a process, idea, or concept increases when you work to help someone else learn it. These activities give you opportunities to develop deeper understanding by "putting yourself in the instructor's shoes."

Data Analysis and Graphing. These activities are designed to give you practice with interpreting graphs, analyzing data, and/or developing graphs from data sets. These skills are integral to both understanding and communicating information in biology.

Name_____   Course/Section_____

Date_____   Professor/TA_____

Activity 2.1 A Quick Review of Elements and Compounds

1. Table 2.1 (page 32) lists the chemical elements that occur naturally in the human body. Similar percentages of these elements are found in most living organisms.

a. In what abiotic (nonlife) chemical forms are these elements often found in nature?	b. In what chemical form(s) do animals need to obtain these elements?	c. In what chemical form(s) do plants need to obtain these elements?

2. A chemical element cannot be broken down to other forms by chemical reactions. Each element has a specific number of protons, neutrons, and electrons.

    a. What is the name of the following element, and how many protons, neutrons, and electrons does it have?

Name	Number of protons	Number of neutrons	Number of electrons

    b. What information do you need to calculate or determine the following?

The atomic number of an element	The mass number of an element	The weight in daltons of one atom of an element

c. What are the atomic number, mass number, and weight in daltons of the element shown in part a?

Atomic number	Mass number	Weight in daltons

3. One mole of an element or compound contains 6.02×10^{23} atoms or molecules of the element or compound. One mole of an element or compound has a mass equal to its mass number (or molecular weight) in grams. For example, 1 mole of hydrogen gas (H_2) contains 6.02×10^{23} molecules and weighs 2 g.

a. What is the weight of 1 mole of pure sodium (Na)?	b. How many molecules of Na are in 1 mole of Na?

c. How would you determine how many grams are in a mole of any chemical element or compound?

4. One atom of Na can combine with one atom of Cl (chlorine) to produce one molecule of NaCl (table salt).

a. If Cl has 17 electrons, 17 protons, and 18 neutrons, what is its mass number?	b. What is the mass number of NaCl?	c. How many grams of NaCl equal a mole of NaCl?

d. If you wanted to combine equal numbers of Na and Cl atoms in a flask, how much Cl would you have to add if you added 23 g of Na? (Include an explanation of the reasoning behind your answer.)

e. To make a one-molar (1 *M*) solution of NaCl, you need to add 1 mole of NaCl to distilled water to make a final volume of 1 L (1,000 ml). A 1 *M* solution is said to have a molarity of 1. If you added 2 moles of NaCl to 1 L of distilled water, you would make a 2 *M* solution and its molarity would equal 2. You make up a 1 *M* solution of NaCl.

How many molecules of NaCl are in the 1 *M* NaCl solution?	How many molecules of NaCl are there per ml of the solution?

f. Next, you divide this 1 *M* solution of NaCl into four separate flasks, putting 250 mL into each flask.

How many grams of NaCl are in each flask?	How many molecules of NaCl are in each flask?	How many molecules of NaCl are there per ml of distilled water?	What is the molarity of NaCl in each of the four flasks?

5. The summary formula for photosynthesis is

$$6\ CO_2 + 6\ H_2O \rightarrow C_6H_{12}O_6 + 6\ O_2$$

a. How many molecules of carbon dioxide and water would a plant have to use to produce three molecules of glucose ($C_6H_{12}O_6$)?	b. How many moles of carbon dioxide and water would a plant have to use to produce 2 mole of glucose?

c. Refer to the summary formula for photosynthesis. If you know the number of molecules or moles of any of the reactants used (or products produced), how would you calculate the number of molecules or moles of all of the other reactants needed and products produced?

6. A biologist places a plant in a closed chamber. A sensor in the chamber maintains the carbon dioxide level at the normal atmospheric concentration of 0.03%. Another sensor allows the biologist to measure the amount of oxygen produced by the plant over time. If the plant produces 0.001 mole of oxygen in an hour, how much carbon dioxide had to be added to the chamber during that hour to maintain the atmospheric concentration of 0.03%?

7. O_2 and NH_3 are both small covalent molecules found in cells. NH_3 is extremely soluble in the aqueous environment of the cell, while O_2 is relatively insoluble. What is the basis for this difference in solubility between the two molecules? In reaching your answer, draw the structures of the molecules as valence shell diagrams (as in Figure 2.12, page 38). Given these diagrams, consider the types of interactions each molecule could have with water.

8. Refer to pages 38–41 of *Biology*, 8th edition, which describe these types of chemical bonds: nonpolar and polar covalent bonds, ionic bonds, hydrogen bonds, and van der Waals interactions.

   The molecule diagrammed here can also be represented by the formula CH_3COOH.

   Explain how you could determine which of the bonds between elements in this molecule are polar or nonpolar covalent bonds, ionic bonds, hydrogen bonds, and van der Waals interactions.

Name_____   Course/Section_____

Date_____   Professor/TA_____

 Activity 3.1 A Quick Review of the Properties of Water

1. Compounds that have the capacity to form hydrogen bonds with water are said to be hydrophilic (water loving). Those without this capacity are hydrophobic (water fearing).

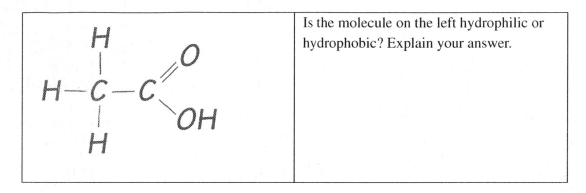

| | Is the molecule on the left hydrophilic or hydrophobic? Explain your answer. |

2. In addition to being polar, water molecules can dissociate into hydronium ions (H_3O^+, often described simply as H^+) and hydroxide ions (OH^-). The concentration of each of these ions in pure water is 10^{-7}. Another way to say this is that the concentration of hydronium ions, or H^+ ions, is one out of every 10 million molecules. Similarly, the concentration of OH^- ions is one in 10 million molecules.

    a. The H^+ ion concentration of a solution can be represented as its pH value. The pH of a solution is defined as the negative $\log_{10}$ of the hydrogen ion concentration. What is the pH of pure water?

    b. Refer to the diagram of the molecule of acetic acid in question 1. The COOH group can ionize to release a H^+ ion into solution. If you add acetic acid to water and raise the concentration of H^+ ions to 10^{-4}, what is the pH of this solution?

3. Life as we know it could not exist without water. All the chemical reactions of life occur in aqueous solution. Water molecules are polar and are capable of forming hydrogen bonds with other polar or charged molecules. As a result, water has the following properties:

    A. H_2O molecules are cohesive; they form hydrogen bonds with each other.
    B. H_2O molecules are adhesive; they form hydrogen bonds with polar surfaces.
    C. Water is a liquid at normal physiological (or body) temperatures.
    D. Water has a high specific heat.
    E. Water has a high heat of vaporization.
    F. Water's greatest density occurs at 4°C.

Explain how these properties of water are related to the phenomena described in parts a–h below. More than one property may be used to explain a given phenomenon.

a. During the winter, air temperatures in the northern United States can remain below 0°C for months; however, the fish and other animals living in the lakes survive.

b. Many substances—for example, salt (NaCl) and sucrose—dissolve quickly in water.

c. When you pour water into a 25-mL graduated cylinder, a meniscus forms at the top of the water column.

d. Sweating and the evaporation of sweat from the body surface help reduce a human's body temperature.

e. A bottle contains a liquid mixture of equal parts water and mineral oil. You shake the bottle vigorously and then set it on the table. Although the law of entropy favors maximum randomness, this mixture separates into layers of oil over water.

f. Water drops that fall on a surface tend to form rounded drops or beads.

g. Water drops that fall on your car tend to bead or round up more after you polish (or wax) the car than before you polished it.

h. If you touch the edge of a paper towel to a drop of colored water, the water will move up into (or be absorbed by) the towel.

Name_____   Course/Section_____

Date_____   Professor/TA_____

Activity 3.1 A Quick Review of the Properties of Water

1. Compounds that have the capacity to form hydrogen bonds with water are said to be hydrophilic (water loving). Those without this capacity are hydrophobic (water fearing).

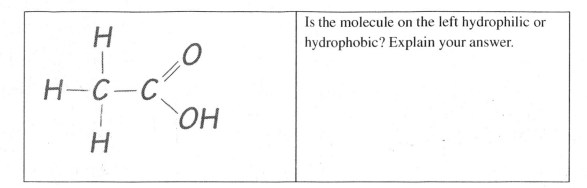

	Is the molecule on the left hydrophilic or hydrophobic? Explain your answer.

2. In addition to being polar, water molecules can dissociate into hydronium ions (H_3O^+, often described simply as H^+) and hydroxide ions (OH^-). The concentration of each of these ions in pure water is 10^{-7}. Another way to say this is that the concentration of hydronium ions, or H^+ ions, is one out of every 10 million molecules. Similarly, the concentration of OH^- ions is one in 10 million molecules.

   a. The H^+ ion concentration of a solution can be represented as its pH value. The pH of a solution is defined as the negative $\log_{10}$ of the hydrogen ion concentration. What is the pH of pure water?

   b. Refer to the diagram of the molecule of acetic acid in question 1. The COOH group can ionize to release a H^+ ion into solution. If you add acetic acid to water and raise the concentration of H^+ ions to 10^{-4}, what is the pH of this solution?

3. Life as we know it could not exist without water. All the chemical reactions of life occur in aqueous solution. Water molecules are polar and are capable of forming hydrogen bonds with other polar or charged molecules. As a result, water has the following properties:
   A. H_2O molecules are cohesive; they form hydrogen bonds with each other.
   B. H_2O molecules are adhesive; they form hydrogen bonds with polar surfaces.
   C. Water is a liquid at normal physiological (or body) temperatures.
   D. Water has a high specific heat.
   E. Water has a high heat of vaporization.
   F. Water's greatest density occurs at 4°C.

   Explain how these properties of water are related to the phenomena described in parts a–h below. More than one property may be used to explain a given phenomenon.

   a. During the winter, air temperatures in the northern United States can remain below 0°C for months; however, the fish and other animals living in the lakes survive.

   b. Many substances—for example, salt (NaCl) and sucrose—dissolve quickly in water.

   c. When you pour water into a 25-mL graduated cylinder, a meniscus forms at the top of the water column.

   d. Sweating and the evaporation of sweat from the body surface help reduce a human's body temperature.

   e. A bottle contains a liquid mixture of equal parts water and mineral oil. You shake the bottle vigorously and then set it on the table. Although the law of entropy favors maximum randomness, this mixture separates into layers of oil over water.

   f. Water drops that fall on a surface tend to form rounded drops or beads.

   g. Water drops that fall on your car tend to bead or round up more after you polish (or wax) the car than before you polished it.

   h. If you touch the edge of a paper towel to a drop of colored water, the water will move up into (or be absorbed by) the towel.

Name_____   Course/Section_____

Date_____   Professor/TA_____

Activity 4.1/5.1 How can you identify organic macromolecules?

Refer to the figure (Some Simple Chemistry) on the next page when doing this activity.

Part A. Answer the questions. Then use your answers to develop simple rules for identifying carbohydrates, lipids, proteins, and nucleic acids.

1. What is the approximate C:H:O ratio in each of the following types of macromolecules?

Carbohydrates	Lipids	Proteins	Nucleic acids

2. Which of the compounds listed in question 1 can often be composed of C, H, and O alone?

3. Which of the compounds can be identified by looking at the C:H:O ratios alone?

4. What other elements are commonly associated with each of these four types of macromolecules?

	Carbohydrates	Lipids	Proteins	Nucleic acids
Always contain P				
Generally contain no P				
Always contain N				
Generally contain no N				
Frequently contain S				
Generally contain no S				

Some Simple Chemistry

Compound	Basic components ➡ Reaction ➡ Product

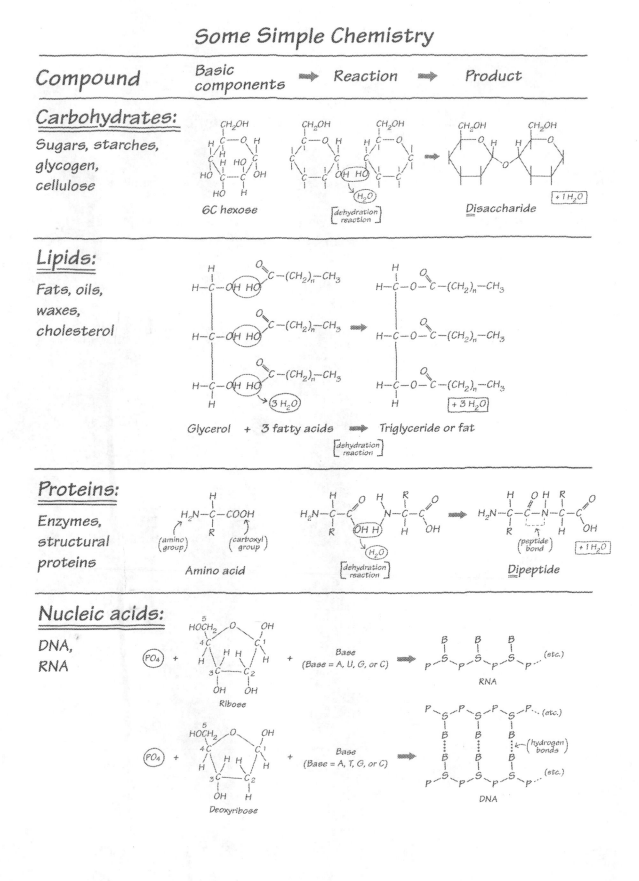

Carbohydrates:

Sugars, starches, glycogen, cellulose

6C hexose
dehydration reaction
Disaccharide
+ 1 H₂O

Lipids:

Fats, oils, waxes, cholesterol

Glycerol + 3 fatty acids ➡ Triglyceride or fat
dehydration reaction
+ 3 H₂O

Proteins:

Enzymes, structural proteins

amino group
carboxyl group
Amino acid
dehydration reaction
peptide bond
Dipeptide
+ 1 H₂O

Nucleic acids:

DNA, RNA

Ribose
Base (Base = A, U, G, or C) ➡ RNA

Deoxyribose
Base (Base = A, T, G, or C) ➡ DNA
hydrogen bonds

5. Functional groups can modify the properties of organic molecules. In the following table, indicate whether each functional group is polar or nonpolar and hydrophobic or hydrophilic. Which of these functional groups are found in proteins and lipids?

Functional group	Polar or nonpolar	Hydrophobic or hydrophilic	Found in all proteins	Found in many proteins	Found in many lipids
—OH					
—CH₂					
—COOH					
—NH₂					
—SH					
—PO₄					

6. You want to use a radioactive tracer that will label only the protein in an RNA virus. Assume the virus is composed of only a protein coat and an RNA core. Which of the following would you use? Be sure to explain your answer.

   a. Radioactive P   b. Radioactive N   c. Radioactive S   d. Radioactive C

7. Closely related macromolecules often have many characteristics in common. For example, they share many of the same chemical elements and functional groups. Therefore, to separate or distinguish closely related macromolecules, you need to determine how they differ and then target or label that difference.

   a. What makes RNA different from DNA?

   b. If you wanted to use a radioactive or fluorescent tag to label only the RNA in a cell and not the DNA, what compound(s) could you label that is/are specific for RNA?

   c. If you wanted to label only the DNA, what compound(s) could you label?

8. Based on your answers to questions 1–7, what simple rule(s) can you use to identify the following macromolecules?

Carbohydrates	
Lipids	
Proteins	
Nucleic acids	
DNA versus RNA	

Part B. Carbohydrate, lipid, protein, or nucleic acid? Name that structure!

Based on the rules you developed in Part A, identify the compounds below (and on the following page) as carbohydrates, lipids, amino acids, polypeptides, or nucleic acids. In addition, indicate whether each is likely to be polar or nonpolar, hydrophilic or hydrophobic.

1)
$$C_{17}H_{35}COOH \; + \quad
\begin{array}{c}
H \\
| \\
H\!-\!O\!-\!\overset{|}{C}\!-\!H \\
| \\
H\!-\!O\!-\!\overset{|}{C}\!-\!H \\
| \\
H\!-\!O\!-\!\overset{|}{C}\!-\!H
\end{array}
\quad \longrightarrow \quad
\begin{array}{c}
H \\
| \\
C_{17}H_{35}COO\!-\!\overset{|}{C}\!-\!H \\
| \\
C_{17}H_{35}COO\!-\!\overset{|}{C}\!-\!H \\
| \\
C_{17}H_{35}COO\!-\!\overset{|}{C}\!-\!H
\end{array}$$

2)
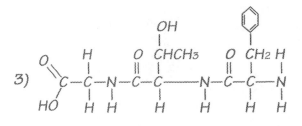

3)

Name_____   Course/Section_____

Part B. *Continued*

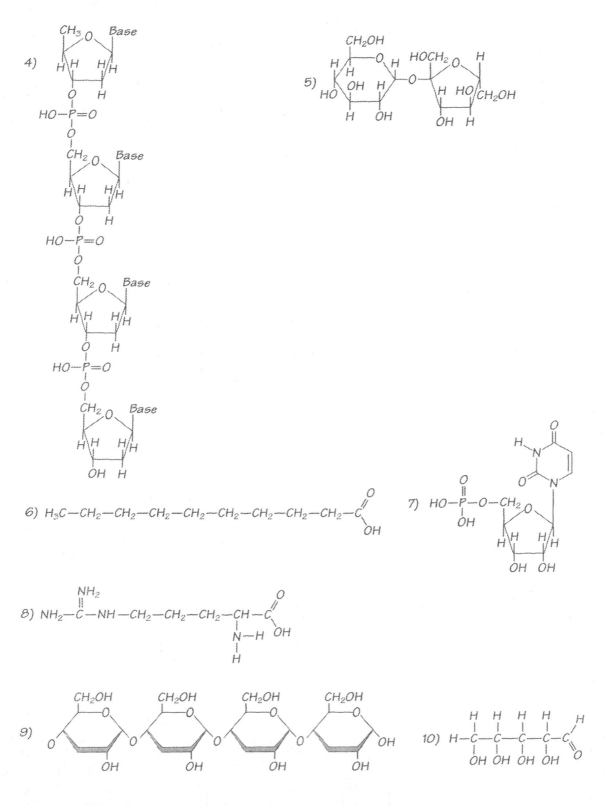

4.1/5.1 Test Your Understanding

A student, Mary, is given four samples and told they are lysine (an amino acid), lactose (a disaccharide), insulin (a protein hormone), and RNA. The samples are in test tubes marked 1, 2, 3, and 4, but Mary doesn't know which compound is in which tube. She is instructed to identify the contents of each tube.

a. In her first test, she tries to hydrolyze a portion of the contents of each tube. Hydrolysis occurs in all tubes except tube 3.

b. In Mary's next test, she finds that tubes 1, 2, and 3 are positive for nitrogen but only tube 2 gives a positive result for the presence of sulfur.

c. The last test Mary performs shows that the compound in tube 1 contains a high percentage of phosphate.

Based on these data, fill in the following table and explain your answers.

Tube number	Contents	Explanation
1		
2		
3		
4		

Name_____   Course/Section_____

Date_____   Professor/TA_____

Activity 4.2/5.2 What predictions can you make about the behavior of organic macromolecules if you know their structure?

1. Twenty amino acids are commonly utilized in the synthesis of proteins. These amino acids differ in the chemical properties of their side chains (also called R groups). What properties does each of the following R groups have? (*Note:* A side chain may display more than one of these properties.)

R group	Basic, acidic, or neutral	Polar or nonpolar	Hydrophilic or hydrophobic
a. CH_2 / CH / CH_3 CH_3			
b. CH_2 / $C=O$ / $-O$			
c. CH_2 / CH_2 / CH_2 / CH_2 / NH_3^+			
d. CH_2 / OH			

2. Polypeptides and proteins are made up of linear sequences of amino acids. In its functional form, each protein has a specific three-dimensional structure or shape. Interactions among the individual amino acids and their side chains play a major role in determining this shape.

   a. How are amino acids linked together to form polypeptides or proteins? What is this type of bond called?

b. Define the four structures of a protein.	c. What kinds of bonds hold each of these structures together?
Primary:	
Secondary:	
Tertiary:	
Quaternary:	

3. Lipids as a group are defined as being hydrophobic, or insoluble in water. As a result, this group includes a fairly wide range of compounds—for example, fats, oils, waxes, and steroids like cholesterol.

   a. How are fatty acids and glycerol linked together to form fats (triglycerides)?

   b. What functions do fats serve in living organisms?

   c. How do phospholipids differ from triglycerides?

   d. What characteristics do phospholipids have that triglycerides do not have?

4.2/5.2 Test Your Understanding

Use your understanding of the chemical characteristics of the four major types of macromolecules in living organisms to predict the outcome of the following experiments. Be sure to explain your reasoning.

Experiment 1: You stir 10 g of glucose and 10 mL of phospholipids in a 500-mL beaker that contains 200 mL of distilled water. Draw a diagram to show where and how the glucose and phospholipids would be distributed after you let the mixture settle for about 30 minutes.

Experiment 2: You repeat Experiment 1 again, but this time you stir 10 g of glucose and 10 mL of phospholipids in a different 500-mL beaker that contains 200 mL of distilled water and 100 mL of oil. Draw a diagram to show where and how the glucose, phospholipids, and oil would be distributed after you let the solution settle for about 30 minutes.

Experiment 3: To completely fill a sealed 500-mL glass container that contains 490 mL of distilled water, you inject 10 mL of phospholipids into it. (A small gasket allows the air to leave as you inject the phospholipids.) You shake this mixture vigorously and then let it settle for an hour or more. Draw a diagram to show how the phospholipids would be distributed in the container.

Experiment 4: A globular protein that is ordinarily found in aqueous solution has these amino acids in its primary structure: glutamic acid, lysine, leucine, and tryptophan. Predict where you would find each amino acid: in the interior portion of the protein (away from water) or on the outside of the protein (facing water). (Refer to Figure 5.17, page 79.)

Experiment 5: Drawn below is part of the tertiary structure of a protein showing the positions of two amino acids (aspartic acid and lysine). Replacing lysine with another amino acid in the protein may change the shape and function of the protein. Replacing lysine with which type(s) of amino acid(s) would lead to the least amount of change in the tertiary structure of this protein? (Refer to Figure 5.17, page 79.)

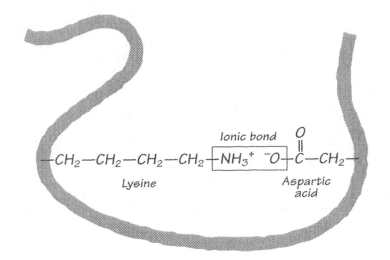

Name_____   Course/Section_____

Date_____   Professor/TA_____

 Activity 6.1 What makes a cell a living organism?

1. Single-celled organisms and individual cells within multicellular organisms can vary
    greatly in appearance as well as in the functions they perform. Nonetheless, each of
    these cells is alive and therefore must have some common characteristics.

a. At a minimum, what structures or components must a cell contain to be alive?	b. What is the function of each structure or component listed in part a?

c. If you consider the types of single-celled organisms that exist today, which, if
    any, have a structure similar to your description in part a?

2. What would you need to add to or change about the cell you described in question 1
    to make it:

a. A eukaryotic animal cell?	b. A eukaryotic plant cell?

3. To get an idea of the different sizes of various cellular components, do the following calculations: Assume that the cell, its nucleus, and a globular protein—for example, an enzyme—are spherical. In addition, assume the diameter of the protein is 5 nm, the diameter of the cell is 100 μm (micrometers), and the diameter of the nucleus is 40 μm.

If you draw the globular protein as a sphere with a diameter of 2 cm (approximately the diameter of a U.S. penny), what size would each of the following measurements of the cell be if drawn to the same scale (5-nm real length = 2 cm)?	
a. The radius of a microtubule (Refer to Table 6.1, page 113, in *Biology*, 8th ed.)	
b. The diameter of the nucleus	
c. The diameter of the cell	
d. The volume ($V = 4/3\ \pi r^3$) of the protein 1 nanometer cubed $(1\ nm^3) = 1.0 \times 10^{-21}$ centimeters cubed (cm^3).	
e. The volume of the nucleus	
f. The volume of the cell	

g. The volume of the Empire State Building is $1.05 \times 106\ m^3$. How many of your scaled nuclei could fit into the Empire State Building? How many of your scaled cells could fit?

h. Do the results of these calculations help you to understand how so much can be going on inside a cell at once? Explain.

Name_____   Course/Section_____

Date_____   Professor/TA_____

Activity 7.1 What controls the movement of materials into and out of the cell?

1. To be alive, most cells must maintain a relatively constant internal environment. To do this, they must be able to control the movement of materials into and out of the cell.

What characteristics of the cell membrane determine what gets into the cell and what doesn't? That is, what determines the permeability of a cell or organelle membrane? To answer these questions, first consider the answers to the following questions:

a. If a cell membrane were composed of only a phospholipid bilayer, what properties would it have?	b. What different roles or functions do membrane proteins serve?	c. Why are some cell types more permeable to a substance (for example, sodium ions) than others?

Using your understanding of the answers in parts a–c, now answer these questions: What characteristics of the cell membrane determine what gets into the cell and what doesn't? That is, what determines the permeability of a cell or organelle membrane?

2. You design an experiment to test the effect(s) various compounds have on the osmotic potential of a model cell. You know that substances dissolved in aqueous or gaseous solutions tend to diffuse from regions of higher concentration to regions of lower concentration.

You fill each of three (20-mL) dialysis bags half full with one of these substances:
- 5% by weight of glucose in distilled water
- 5% by weight of egg albumin (protein) in distilled water
- 5% by weight of glass bead (one glass bead) in distilled water

The dialysis bag is permeable to water but impermeable to glucose, albumin, and glass bead.

a. If the final weight of each bag is 10 g, how many grams of glucose, albumin, and glass bead were added to each bag?

b. The molecular weight of the protein is about 45 kilodaltons, and the molecular weight of glucose is about 180 daltons. How can you estimate the number of molecules of glucose in the 5% solution compared to the number of albumin molecules in its 5% solution?

c. You put the dialysis bags into three separate flasks of distilled water. After 2 hours, you remove the bags and record these weights:

Dialysis bag	Weight
Glucose	13.2 g
Albumin	10.1 g
Glass bead	10.0 g

How do you explain these results? (*Hint:* Consider the surface-area-to-volume ratio of each of the three substances and review pages 50 and 51 of *Biology,* 8th edition.)

d. What results would you predict if you set up a similar experiment but used 5% glucose and 5% sucrose?

Name_____   Course/Section_____

Date_____   Professor/TA_____

 Activity 7.2 How is the structure of a cell membrane related to its function?

Membranes compartmentalize the different functions of living cells. The cell membrane is a barrier between the cell or organism and its environment. Similarly, within the cell, membranes of organelles separate the different reactions of metabolism from each other.

Use the supplies provided in class or devise your own at home to develop a model of a cell membrane. Developing models of systems can help you understand not only their overall structure but also their function(s).

Building the Model

- Include in the membrane the phospholipid bilayer (phosphate heads and fatty acid tails) as well as the integral proteins.
- Design integral proteins that serve the functions of facilitated diffusion and active transport.
- Indicate how the various types of integral proteins might differ in structure and operation.

Use the understanding you gain from your model to answer the questions on the next pages.

1. Substances can move across the membrane via simple diffusion, facilitated diffusion, or active transport.

	a. Where does it occur in membrane?	b. Does it require transport protein?	c. Does it require input of energy?
Simple diffusion			
Facilitated diffusion			
Active transport			

d. What functions might each of the three types of diffusion serve in an independent cell such as a *Paramecium* or an amoeba?

e. What functions might each of the three types of diffusion serve in a multicellular organism—for example, a human or a tree?

2. What would you need to observe or measure to determine whether a substance was moved across a membrane via each type of diffusion?

Simple diffusion	Facilitated diffusion	Active transport

3. The ratios of saturated to unsaturated phospholipids in an organism's membranes can change in response to changes in environmental conditions.

    a. How do the properties of a membrane that contains a low percentage of unsaturated phospholipids compare with those of a membrane that contains a high percentage of unsaturated phospholipids?

    b. Considering what you know about the properties of saturated and unsaturated fatty acids, would you expect an amoeba that lives in a pond in a cold northern climate to have a higher or lower percentage of saturated fatty acids in its membranes during the summer as compared to the winter? Explain your answer.

4. A fish is removed from a contaminated lake. You determine that a particular toxin (X) is present in its cells at concentration X = 1,500 μg/L. You place the fish in a tank full of clean water (X = 0 μg/L), and measure the toxin concentration in the fish cells each day for the next 10 days.

a. On the graphs below, predict how the toxin concentrations in the fish and in the water will change over time if:

  i. the toxin is water soluble

  ii. the toxin is fat soluble

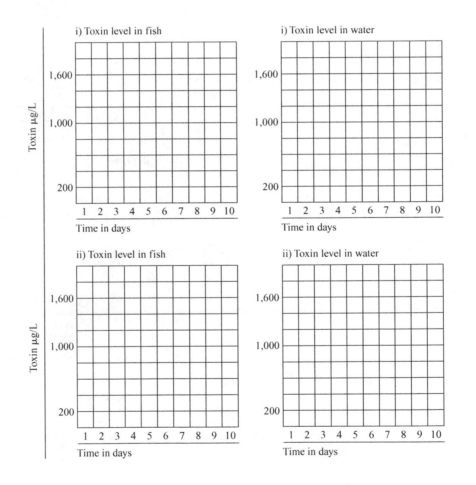

b. After making your hypothesis, you test it by measuring the toxin levels in the fish at various times during its 10 days in the tank. You observe that the level of toxin in the fish drops from 1,500 µg/L to 750 µg/L and then stabilizes at 750 µg/L. You test the water in the tank and find that after it stabilizes, toxin is present in the water at concentration 750 µg/L also.

- Which of your predictions fits these data?
- Which of the following processes is most likely eliminating the toxin from the fish?

    i. Passive transport

    ii. First active, then passive transport

    iii. First passive, then active transport

    iv. Active transport

c. Given the situation in part b, what should you do, in the short term, to continue to reduce the toxin level in the fish below 750 µg/L?

5. A particular amino acid is transported from the extracellular medium against its concentration gradient. The integral membrane protein that transports the amino acid also binds and transports Na^+. Using your model of the cell membrane, develop a transport mechanism that will permit the amino acid uptake to be coupled to the Na^+ transport so that the amino acid's entry is linked only indirectly to ATP hydrolysis.

Name_____   Course/Section_____

Date_____   Professor/TA_____

Activity 8.1 What factors affect chemical reactions in cells?

Construct a concept map of general metabolism using the terms in the list below. Keep in mind that there are many ways to construct a concept map.

- Begin by writing each term on a separate sticky note or piece of paper.
- Then organize the terms into a map that indicates how the terms are associated or related.
- Draw lines between terms and add action phrases to the lines to indicate how the terms are related.
- If you are doing this activity in small groups in class, explain your map to another group when you finish it.

Here is an example:

Terms

peptide bonds

proteins

α helix

primary structure

secondary structure

tertiary structure

β pleated sheet

R groups

hydrogen bonds

substrate or reactant
   (ligand)

activation energy

ΔG / free energy

endergonic

exergonic

enzymes

catalysts

competitive inhibitor

noncompetitive inhibitor

active site

product

allosteric regulation

activator

four-step enzyme-mediated
   reaction sequence
   or metabolic pathway
   $(A \rightarrow B \rightarrow C \rightarrow D)$

intermediate compound

end product

feedback inhibition

Use the understanding you gained from doing the concept map to answer the questions.

1. Reduced organic compounds tend to contain stored energy in C—H bonds. As a general rule, the greater the number of C—H bonds, the greater the amount of potential energy stored in the molecule. Answer each question in the chart as it relates to the two reactions shown at the top. Be sure to explain the reasoning behind your answers.

	Reaction 1: $CH_4 + 2 O_2 \rightarrow H_2O + CO_2$ (methane)	Reaction 2: $6 CO_2 + 6 H_2O \rightarrow C_6H_{12}O_6 + 6 O_2$
a. Is the reaction exergonic or endergonic?		
b. Is the reaction spontaneous?		
c. Is the reaction anabolic or catabolic?		
d. Is ΔG (the change in free energy) positive or negative?		

2. All metabolic reactions in living organisms are enzyme mediated. Each enzyme is specific for one (or only a very few similar types of) reaction. Given this, there are approximately as many different kinds of enzymes as there are reactions.

   a. What characteristics do all enzymes share?

   b. What characteristics can differ among enzymes?

3. How can enzyme function be mediated or modified? To answer, complete a and b below.

a. What factors can modify enzyme function?	b. What effect(s) can each of these factors have on enzyme function?

c. What role(s) can modification of enzyme function play in the cell?

Activity 8.2 How can changes in experimental conditions affect enzyme-mediated reactions?

1. You set up a series of experiments to monitor the rates of a reaction. The reaction is an enzyme-mediated reaction in which A → B + C. For each experiment in this series, you continuously add the reactant A and monitor its concentration so that the amount of A remains constant over time.

    For each group of experiments, explain how the differences in experimental conditions could affect the reaction.

    a. You compare two side-by-side experiments. In experiment 1, you use *X* amount of the enzyme. In experiment 2, you use 2*X* amount of the same enzyme.

    b. You compare two side-by-side experiments. In both you use equal amounts of the enzyme. In experiment 3, you allow the products to accumulate over time. In experiment 4, you remove the products from the system as they are produced.

    c. In the next two experiments, you use equal amounts of the enzyme. You run experiment 5 at 20°C and experiment 6 at 25°C.

    d. In two final experiments, you use equal amounts of the enzyme. You run experiment 7 at pH 6 and experiment 8 at pH 8.

2. Enzyme function can be inhibited or regulated by the presence of chemicals that mimic either the reactants or the products.

   a. How do competitive and noncompetitive inhibition of an enzyme differ?

   b. What are allosteric enzymes? What function(s) can they serve in reaction sequences?

3. An enzyme catalyzes the reaction X → Y + Z. In a series of experiments, it was found that substance A inhibits the enzyme.

   • When the concentration of X is high and A is low, the reaction proceeds rapidly and Y and Z are formed.
   • As the concentration of A increases, the reaction slows regardless of whether X is present in high or low concentration.
   • If the concentration of A is high (relative to X), the reaction stops.
   • If the concentration of A again decreases, the reaction will ultimately resume.

   What type of enzyme regulation is described here? Explain or justify your answer.

4. In an enzymatic pathway, A, B, and C are intermediates required to make D, and 1, 2, and 3 are enzymes that catalyze the designated reactions:

$$\begin{array}{ccccccc} & 1 & & 2 & & 3 \\ A & \to & B & \to & C & \to & D \\ \downarrow & & & & & & \\ E & & & & & & \end{array}$$

   This is analogous to what happens in a factory. In a leather goods factory, for example, the leather (A) is cut (1) into the parts needed for shoes (B). The shoe parts are sewn (2) together (making C), and C is packaged (3) for shipping as D. Now shoe sales are dropping and backpack sales (E) are increasing. As a result, the manager of the factory decides to switch production from shoes to leather backpacks (E).

   a. Where should the shoe-making process be shut down: step 1, 2, or 3? Explain.

b. In a cell, if an excess of a chemical product D arises, where should this synthetic pathway be shut down in the cell? Explain your reasoning.

c. What type(s) of enzyme regulation is/are most likely to occur in the cell in this type of feedback system? Explain your reasoning.

Name_____   Course/Section_____

Date_____   Professor/TA_____

Activity 9.1 A Quick Review of Energy Transformations

Review Chapter 8 and pages 162–164 of Chapter 9 in *Biology*, 8th edition. Then complete the discussion by supplying or choosing the appropriate terms.

To maintain life, organisms must be able to convert energy from one form to another. For example, in the process of photosynthesis, algae, plants, and photosynthetic prokaryotes use the energy from sunlight to convert carbon dioxide and water to glucose and oxygen (a waste product).

The summary reaction for photosynthesis can be written as

$$6\,CO_2 + 6\,H_2O \rightarrow C_6H_{12}O_6 + 6\,O_2$$

This type of reaction is an oxidation-reduction (or redox) reaction. This reaction is also **[anabolic/catabolic]** and **[endergonic/exergonic]**.

In redox reactions, _____ (and associated H^+ ions) are transferred from one compound or element to another. If one compound or element "loses" _____ and becomes oxidized, another must "gain" _____ and become reduced. For example, in photosynthesis, water becomes **[oxidized/reduced]** (to O_2) and the _____ (and associated H^+ ions) it loses in the process **[oxidize/reduce]** CO_2 to glucose.

[Anabolic/Catabolic] reactions "build" more complex molecules from simpler ones. To do this they require energy input. Reactions that require the input of energy are termed **[endergonic/exergonic]** reactions.

The reactions involved in aerobic respiration are also redox reactions:

$$C_6H_{12}O_6 + 6\,O_2 \rightarrow 6\,CO_2 + 6\,H_2O$$

In this set of reactions, however, more complex molecules are "broken down" into simpler ones. Glucose is broken down or becomes **[oxidized/reduced]** (to CO_2), and the oxygen becomes **[oxidized/reduced]** (to water).

[Anabolic/Catabolic] reactions break down more complex molecules into simpler ones and in the process release energy. Reactions that release energy that can be used to do work are **[endergonic/exergonic]**. Therefore, aerobic respiration is a(n) **[anabolic/catabolic]** process and is **[endergonic/exergonic]**.

[**Endergonic/exergonic**] reactions are also said to be spontaneous reactions. Does this mean that if we don't keep glucose in tightly sealed containers it will spontaneously interact with atmospheric oxygen and turn into carbon dioxide and water? The answer is obviously no.

Spontaneous reactions rarely occur "spontaneously" because all chemical reactions, even those that release energy, require some addition of energy—the energy of activation—before they can occur. One way of supplying this energy is to add heat. An example is heating a marshmallow over a flame or campfire. When enough heat is added to reach (or overcome) the activation energy, the sugar in the marshmallow reacts by oxidizing. (Burning is a form of oxidation.) The marshmallow will continue to burn even if you remove it from the campfire. As the marshmallow burns, carbon dioxide and water are formed as products of the reaction, and the energy that was stored in the bonds of the sugar is released as heat.

If our cells used heat to overcome activation energies in metabolism, they would probably burn up like the marshmallow did. Instead, living systems use protein catalysts or enzymes to lower the energy of activation without adding heat. In addition, the metabolic breakdown of sugars is carried out in a controlled series of reactions. At each step or reaction in the sequence, a small amount of the total energy is released. Some of this energy is still lost as heat. The rest is converted to other forms that can be used in the cell to drive or fuel coupled endergonic reactions or to make ATP.

Name_____   Course/Section_____

Date_____   Professor/TA_____

Activity 9.2 Modeling cellular respiration: How can cells convert the energy in glucose to ATP?

Using your textbook, lecture notes, and the materials available in class (or those you devise at home), model both fermentation (an anaerobic process) and cellular respiration (an aerobic process) as they occur in a plant or animal cell. Each model should include a dynamic (working or active) representation of the events that occur in glycolysis.

Building the Model

- Use chalk on a tabletop or a marker on a large sheet of paper to draw the cell membrane and the mitochondrial membranes.
- Use playdough or cutout pieces of paper to represent the molecules, ions, and membrane transporters or pumps.
- Use the pieces you assembled to model the processes of fermentation and aerobic respiration. Develop a dynamic (claymation-type) model that allows you to manipulate or move glucose and its breakdown products through the various steps of both fermentation and aerobic respiration.
- When you feel you have developed a good working model, demonstrate and explain it to another student.

Be sure your model of **fermentation** includes and explains the actions and roles of the following:

glycolysis	ADP
cytoplasm	P_i
electrons	ATP
protons	pyruvate
glucose	ethyl alcohol (or lactic acid)
NAD^+	substrate-level phosphorylation
NADH	

Be sure your model of **cellular respiration** includes and explains the actions and roles of the following:

glucose	electron transport chain
oxygen	mitochondria
carbon dioxide	inner mitochondrial membrane
pyruvate	outer mitochondrial membrane
acetyl CoA	H^+
NAD^+	electrons (e^-)
NADH	chemiosmosis
FAD	ATP synthase (proton pumps)
$FADH_2$	cristae
ADP	proton gradients
Ⓟᵢ	oxidative phosphorylation
ATP	substrate-level phosphorylation
water	oxidative phosphorylation

Use your models to answer the questions.

1. The summary formula for cellular respiration is

$$C_6H_{12}O_6 + 6\, O_2 \rightarrow 6\, CO_2 + 6\, H_2O + Energy$$

a. At what stage(s) in the overall process is each of the reactants used?		b. At what stage(s) in the overall process is each of the products produced?		
$C_6H_{12}O_6$ + $6\, O_2$ $\rightarrow$		$6\, CO_2$ + $6\, H_2O$ +		Energy

2. In cellular respiration, the oxidation of glucose is carried out in a controlled series of reactions. At each step or reaction in the sequence, a small amount of the total energy is released. Some of this energy is lost as heat. The rest is converted to other forms that can be used by the cell to drive or fuel coupled endergonic reactions or to make ATP.

a. What is/are the overall function(s) of glycolysis?	b. What is/are the overall function(s) of the Krebs cycle?	c. What is/are the overall function(s) of oxidative phosphorylation?

3. Are the compounds listed here *used* or *produced* in:	Glycolysis?	The Krebs cycle?	Oxidative phosphorylation?
Glucose			
O_2			
CO_2			
H_2O			
ATP			
ADP + (P_i)			
NADH			
NAD^+			

4. The cell's supply of ADP, P_i, and NAD^+ is finite (limited). What happens to cellular respiration when all of the cell's NAD^+ has been converted to NADH?

5. If the Krebs cycle does not require oxygen, why does cellular respiration stop after glycolysis when no oxygen is present?

6. Many organisms can withstand periods of oxygen debt (anaerobic conditions). Yeast undergoing oxygen debt converts pyruvic acid to ethanol and carbon dioxide. Animals undergoing oxygen debt convert pyruvic acid to lactic acid. Pyruvic acid is fairly nontoxic in even high concentrations. Both ethanol and lactic acid are toxic in even moderate concentrations. Explain why this conversion occurs in organisms.

7. How efficient is fermentation? How efficient is cellular respiration? Remember that efficiency is the amount of useful energy (as ATP) gained during the process divided by the total amount of energy available in glucose. Use 686 kcal as the total energy available in 1 mole of glucose and 8 kcal as the energy available in 1 mol of ATP.

Efficiency of fermentation	Efficiency of aerobic respiration

8. a. Why can't cells store large quantities of ATP? (*Hint:* Consider both the chemical stability of the molecule and the cell's osmotic potential.)

   b. Given that cells can't store ATP for long periods of time, how do they store energy?

   c. What are the advantages of storing energy in these alternative forms?

9. To make a 5 *M* solution of hydrochloric acid, we add 400 mL of 12.5 *M* hydrochloric acid to 600 mL of distilled water. Before we add the acid, however, we place the flask containing the distilled water into the sink because this solution can heat up so rapidly that the flask breaks. How is this reaction similar to what happens in chemiosmosis? How is it different?

a. Similarities	b. Differences

1. If it takes 1,000 g of glucose to grow 10 g of an anaerobic bacterium, how many grams of glucose would it take to grow 10 g of that same bacterium if it was respiring aerobically? Estimate your answer. For example, if it takes X amount of glucose to grow 10 g of anaerobic bacteria, what factor would you have to multiply or divide X by to grow 10 g of the same bacterium aerobically? Explain how you arrived at your answer.

2. Mitochondria isolated from liver cells can be used to study the rate of electron transport in response to a variety of chemicals. The rate of electron transport is measured as the rate of disappearance of O_2 from the solution using an oxygen-sensitive electrode. How can we justify using the disappearance of oxygen from the solution as a measure of electron transport?

3. Humans oxidize glucose in the presence of oxygen. For each mole of glucose oxidized, about 686 kcal of energy is released. This is true whether the mole of glucose is oxidized in human cells or burned in the air. A calorie is the amount of energy required to raise the temperature of 1 g of water by 1°C; 686 kcal = 686,000 calories. The average human requires about 2,000 kcal of energy per day, which is equivalent to about 3 mol of glucose per day. Given this, why don't humans spontaneously combust?

4. A gene has recently been identified that encodes for a protein that increases longevity in mice. To function in increasing longevity, this gene requires a high ratio of NAD^+/NADH. Researchers have used this as evidence in support of a "caloric restriction" hypothesis for longevity—that a decrease in total calorie intake increases longevity. How does the requirement for a high NAD^+/NADH ratio support the caloric restriction hypothesis?

5. An active college-age athlete can burn more than 3,000 kcal/day in exercise.
   a. If conversion of one mole of ATP to ADP + P_i releases about 7.3 kcal, roughly speaking, how many moles of ATP need to be produced per day in order for this energy need to be met?
   b. If the molecular weight of ATP is 573, how much would the required ATP weigh in kilograms?
   c. Explain these results.

Name_____   Course/Section_____

Date_____   Professor/TA_____

Activity 10.1 Modeling photosynthesis: How can cells use the sun's energy to convert carbon dioxide and water into glucose?

Activity 10.1 is designed to help you understand:

1. The roles photosystems I and II and the Calvin cycle play in photosynthesis, and
2. How and why C_4 and CAM photosynthesis differ from C_3 photosynthesis.

Using your textbook, lecture notes, and the materials available in class (or those you devise at home), model photosynthesis as it occurs in a plant cell.

Your model should be a dynamic (working or active) representation of the events that occur in the various phases of C_3 photosynthesis.

Building the Model

- Use chalk on a tabletop or a marker on a large sheet of paper to draw the cell membrane and the chloroplast membranes.
- Use playdough or cutout pieces of paper to represent the molecules, ions, and membrane transporters or pumps.
- Use the pieces you assembled to model the processes involved in C_3 photosynthesis. Develop a dynamic (claymation-type) model that allows you to manipulate or move carbon dioxide and water and its breakdown products through the various steps of the process.
- When you feel you have developed a good working model, demonstrate and explain it to another student or to your instructor.

Your model of C_3 photosynthesis should include what occurs in photosystems I and II and in the Calvin cycle. For **photosystems I and II**, be sure your model includes and explains the roles of the following:

$NADP^+$	ATP	chemiosmosis
NADPH	water and oxygen	ATP synthase
ADP	H^+	e^- carriers in thylakoid
P_i	e^-	membranes

Also indicate where in the plant cell each item is required or produced.

For the **Calvin cycle,** be sure your model includes and explains the roles of the following:

glucose   NADPH

C_3 or 3C sugars   ATP

carbon dioxide

Also indicate where in the plant cell each item is required or produced.

After you've modeled C_3 photosynthesis, indicate how the system would be altered for C_4 and CAM photosynthesis.

- Indicate where in the cells of the leaf PEP carboxylase exists and how it reacts to capture CO_2. Be sure to indicate the fate of the captured CO_2.
- Do the same for PEP carboxylase in CAM plants.

Use your model and the information in Chapter 10 of *Biology,* 8th edition, to answer the questions.

1. The various reactions in photosynthesis are spatially segregated from each other within the chloroplast. Draw a simplified diagram of a chloroplast and include these parts: outer membrane, grana, thylakoid, lumen, stroma/matrix.

a. Where in the chloroplast do the light reactions occur?	
b. Where in the chloroplast is the chemiosmotic gradient developed?	
c. Where in the chloroplast does the Calvin cycle occur?	

Name_____   Course/Section_____

2. In photosynthesis, the reduction of carbon dioxide to form glucose is carried out in a controlled series of reactions. In general, each step or reaction in the sequence requires the input of energy. The sun is the ultimate source of this energy.

a. What is/are the overall function(s) of photosystem I?	b. What is/are the overall function(s) of photosystem II?	c. What is/are the overall function(s) of the Calvin cycle?

3. Are the compounds listed here *used* or *produced* in:	Photosystem I?	Photosystem II?	The Calvin cycle?
Glucose			
O_2			
CO_2			
H_2O			
ATP			
ADP + P_i			
NADPH			
$NADP^+$			

4. Which light reaction system (cyclic or noncyclic) would a chloroplast use in each situation?

a. Plenty of light is available, but the cell contains little $NADP^+$.	b. There is plenty of light, and the cell contains a high concentration of $NADP^+$.

5. All living organisms require a constant supply of ATP to maintain life. If no light is available, how can a plant make ATP?

10.1 Test Your Understanding

Chloroplast thylakoids can be isolated and purified for biochemical experiments. Shown below is an experiment in which pH was measured in a suspension of isolated thylakoids before and after light illumination (first arrow). At the time indicated by the second arrow, a chemical compound was added to the thylakoids. Examine these data and address the following questions.

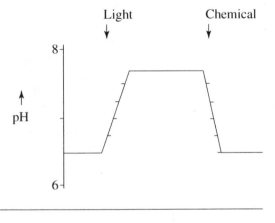

Time (minutes) ➛

a. Based on your understanding of the function of the chloroplasts, why does turning on the light cause the pH in the solution outside the thylakoids to increase?

b. Given the response, the chemical added was probably an inhibitor of:

   i. oxidative phosphorylation

   ii. ATP synthase

   iii. NADPH breakdown

   iv. Electron transport chain between photosystems I and II

   v. Rubisco

Activity 10.1

Name_____   Course/Section_____

Date_____   Professor/TA_____

Activity 10.2 How do C₃, C₄, and CAM photosynthesis compare?

1. Carbon dioxide enters plant leaves through the stomata, while oxygen (the photosynthetic waste product) and water from the leaves exit through the stomata. Plants must constantly balance both water loss and energy gain (as photosynthesis). This has led to the evolution of various modifications of C_3 photosynthesis.

	C_3	C_4	CAM
Draw simplified diagrams of the cross sections of a leaf from a C_3, a C_4, and a CAM plant.			
a. How are the leaves similar?			
b. How are the leaves different?			
c. How and when does carbon dioxide get into each leaf?			
d. Which enzyme(s) (1) capture carbon dioxide and (2) carry it to the Calvin cycle?	(1) (2)	(1) (2)	(1) (2)

e. What makes C_4 photosynthesis more efficient than C_3 photosynthesis in tropical climates?

f. How is CAM photosynthesis advantageous in desert climates?

2. Photosynthesis evolved very early in Earth's history. Central to the evolution of photosynthesis was the evolution of the enzyme rubisco (an abbreviation for ribulose bisphosphate carboxylase oxidase). To the best of our knowledge, all photosynthetic plants use rubisco. Rubisco's function is to supply carbon dioxide to the Calvin cycle; however, it does this only if the ratio of carbon dioxide to oxygen is relatively high. (For comparison, a relatively high ratio of carbon dioxide to oxygen is 0.03% carbon dioxide to 20% oxygen.) When the carbon-dioxide-to-oxygen ratio becomes low, the role of rubisco switches and it catalyzes photorespiration, the breakdown of glucose to carbon dioxide and water.

a. Why could we call photorespiration a "mistake" in the functioning of the cell?

b. Rubisco is thought to have evolved when Earth had a reducing atmosphere. How does this help explain the photorespiration "mistake?"

10.2 Test Your Understanding

The metabolic pathways of organisms living today evolved over a long period of time—undoubtedly in a stepwise fashion because of their complexity. Put the following processes in the order in which they might have evolved, and give a short explanation for your arrangement.

_____ Krebs cycle
_____ Electron transport
_____ Glycolysis
_____ Photosynthesis

Name_____   Course/Section_____

Date_____   Professor/TA_____

 ## Activity 11.1 How are chemical signals translated into cellular responses?

Chapter 11 in *Biology,* 8th edition, describes at least four kinds of signal receptors. Three of these—G-protein-linked receptors, tyrosine-kinase receptors, and ion-channel receptors—are plasma membrane proteins. Protein receptors found in the cytoplasm, or nucleus, of the cell are the fourth type. Some signals (for example, a protein hormone) interact with signal receptors in the cell membrane to initiate the process of signal transduction. This often involves changes in a series of different relay molecules in a signal-transduction pathway. Ultimately, the transduced signal initiates an intracellular response. Other types of signals (for example, steroid hormones) can diffuse through the cell membrane and interact with intracellular receptors. For example, testosterone interacts with its receptor in the cell's cytoplasm, enters the nucleus, and causes the transcription of specific genes.

To help you understand how signal transduction occurs in cells, develop dynamic (claymation-type) models of both a G-protein receptor system and a tyrosine-kinase receptor system. Use playdough or cutout pieces of paper to represent all the structural components and molecules listed here under each system.

G-Protein Receptor System
signal protein

G-protein-linked receptor

plasma membrane

inactive and active G protein

GTP and GDP

inactive and active enzyme

signal-transduction pathway

Tyrosine-Kinase Receptor System
signal protein

tyrosine-kinase receptor

plasma membrane

inactive and active relay proteins

ATP and ADP

signal-transduction pathway

Use your models to show how signal reception by each of the systems can lead to the release of Ca^+ from the endoplasmic reticulum. Demonstrate and explain your models to another student group or to your instructor.

Then use your models to answer the questions on the next page.

1. How are these two systems similar? Consider both structural similarities and similarities in how the systems function.

2. How are the two systems different? Consider both structural differences and differences in how the systems function.

3. Both systems can generate elaborate multistep signal-transduction pathways. These pathways can greatly amplify the cell's response to a signal; the more steps in the pathway, the greater the amplification of the signal. Explain how this amplification can occur. (Review Figure 11.15, page 219, in *Biology*, 8th edition.)

11.1 Test Your Understanding

Humans have the ability to detect and recognize many different aromatic chemicals by smell. Many of these chemicals are present in concentrations less than 1 ppm (part per million) in the air. For example, the majority of humans can detect and recognize chlorine at a concentration of about 0.3 ppm.

a. What characteristics of olfactory (smell) receptors would you look for or propose to explain this ability?

b. Dogs are known to have a much better sense of smell than humans. Given this, what differences may exist in their olfactory system (as compared to humans)?

Name_____   Course/Section_____

Date_____   Professor/TA_____

Activity 12.1 What is mitosis?

What is mitosis?

1. What is the overall purpose of mitosis?

2. In what types of organism(s) and cells does mitosis occur?

3. What type of cell division occurs in bacteria?

What are the stages of mitosis?

4. The fruit fly, *Drosophila melanogaster*, has a total of eight chromosomes (four pairs) in each of its somatic cells. Somatic cells are all cells of the body except those that will divide to form the gametes (ova or sperm). Review the events that occur in the various stages of mitosis.

Keep in mind that the stages of cell division were first recognized from an examination of fixed slides of tissues undergoing division. On fixed slides, cells are captured or frozen at particular points in the division cycle. Using these static slides, early microscopists identified specific arrangements or patterns of chromosomes that occurred at various stages of the cycle and gave these stages names (interphase, prophase, and so on). Later work using time-lapse photography made it clear that mitosis is a continuous process. Once division begins, the chromosomes move fluidly from one phase to the next.

Assume you are a microscopist viewing fruit fly cells that are undergoing mitosis. Within each of the circles (which represent cell membranes) on the following page, draw what you would expect to see if you were looking at a cell in the stage of mitosis indicated. If no circle is present, draw what you would expect to see at the given stage.

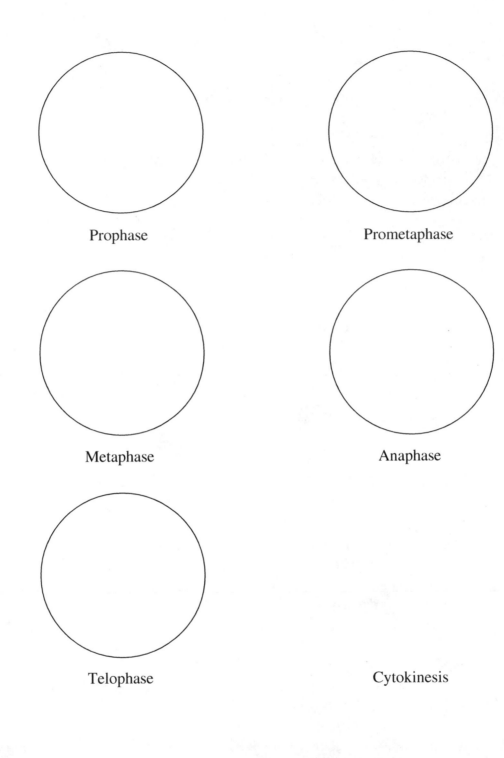

Prophase

Prometaphase

Metaphase

Anaphase

Telophase

Cytokinesis

Daughter cells in
interphase

Name_____   Course/Section_____

What are the products of mitosis?

5. How many cells are produced at the end of a single mitotic division?

6. How many different kinds of cells are produced at the end of a single mitotic division?

7. Six centromeres are observed in a prophase cell from another species of insect.

a. How many pairs of chromosomes does this organism contain?		
b. For each stage of mitosis, indicate the number of centromeres you would expect to find and the number of copies of chromosomes attached to each centromere.		
Stage of mitosis:	Number of centromeres visible per cell	Number of chromosome copies attached to each centromere
Prophase		
Anaphase		

12.1 Test Your Understanding

Haplopappus is an annual flowering plant that grows in deserts. It is of interest because its 2n number is only four.

a. This means that cells in the vegetative parts of the plant that are not undergoing mitosis have how many DNA molecules in their nuclei?

b. During metaphase of mitosis, how many DNA molecules would be in the nucleus?

Name_____   Course/Section_____

Date_____   Professor/TA_____

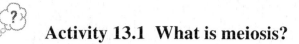 **Activity 13.1 What is meiosis?**

What is meiosis?

1. What is the overall purpose of meiosis?

2. In what types of organism(s) and cells does meiosis occur?

What are the stages of meiosis?

3. The fruit fly, *Drosophila melanogaster*, has a total of eight chromosomes (four pairs) in each of its somatic cells. Somatic cells are all cells of the body except those that will divide to form the gametes (ova or sperm). Review the events that occur in the various stages of meiosis.

Keep in mind that the stages of cell division were first recognized from an examination of fixed slides of tissues undergoing division. On fixed slides, cells are captured or frozen at particular points in the division cycle. Using these static slides, early microscopists identified specific arrangements or patterns of chromosomes that occurred at various stages of the cycle and gave these stages names (interphase, prophase I, and so on). Later work using time-lapse photography made it clear that meiosis is a continuous process. Once division begins, the chromosomes move fluidly from one phase to the next.

Assume you are a microscopist viewing fruit fly cells that are undergoing meiosis. Within each of the circles (which represent cell membranes) on the next pages, draw what you would expect to see if you were looking at a cell in the stage of meiosis indicated. If no circle is present, draw what you would expect to see at the given stage.

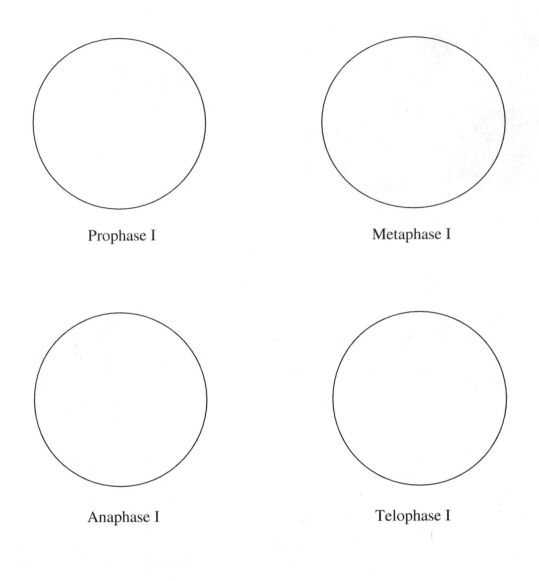

Prophase I

Metaphase I

Anaphase I

Telophase I

Cytokinesis

Daughter cells

Follow one daughter cell through meiosis II.

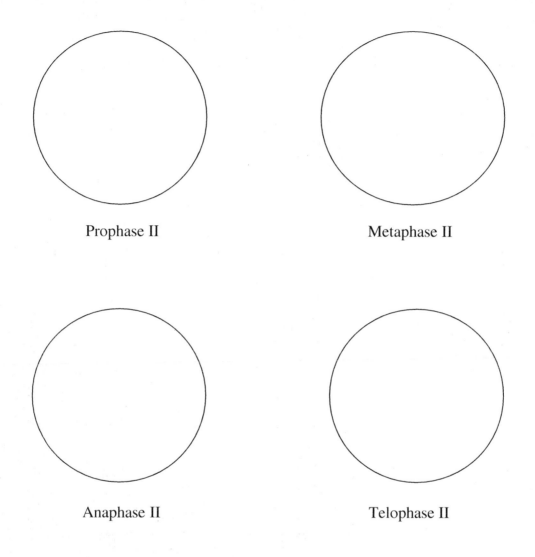

Prophase II   Metaphase II

Anaphase II   Telophase II

Cytokinesis   Daughter cells

What are the products of meiosis?

4. Consider a single cell going through meiosis.

   a. How many cells are produced at the end of meiosis?

   b. How many chromosomes, and which chromosomes, does each of the daughter cells contain?

5. Six centromeres are observed in a prophase I cell from another species of insect.

a. How many pairs of chromosomes does this organism contain?		
b. For each stage of meiosis, indicate the number of centromeres you would expect to find and the number of copies of chromosomes attached to each centromere.		
Stage of meiosis:	Number of centromeres visible per cell	Number of chromosome copies attached to each centromere
Anaphase I		
Prophase II		

13.1 Test Your Understanding

Nondisjunction of sex chromosomes during human gamete formation may lead to individuals with sex chromosome trisomy. An individual with the sex chromosome trisomy of XXY may have resulted from nondisjunction occurring in (circle T if true, F if false):

T/F 1. meiosis I in the father's sperm production

T/F 2. meiosis II in the father's sperm production

T/F 3. meiosis I in the mother's egg production

T/F 4. meiosis II in the mother's egg production

Name_____   Course/Section_____

Date_____   Professor/TA_____

Activity 13.2 How do mitosis and meiosis differ?

Review the processes of mitosis and meiosis in Chapters 12 and 13 of *Biology*, 8th edition, then fill in the chart. Keep in mind that the stages of cell division were first recognized from an examination of fixed slides of tissues undergoing division. On fixed slides, cells are captured or frozen at particular points in the division cycle. Using these static slides, early microscopists identified specific arrangements or patterns of chromosomes that occurred at various stages of the cycle and gave these stages names (interphase, prophase, and so on). Later work using time-lapse photography made it clear that mitosis and meiosis are continuous processes. Once division begins, the chromosomes move fluidly from one phase to the next.

1. What events occur during each phase of mitosis and meiosis?

	Interphase	Prophase	Metaphase	Anaphase	Telophase and cytokinesis
Mitosis	For example: G_1—cell growth S—DNA duplication G_2—cell growth		For example: *Duplicated chromosomes, each with two sister chromatids, line up independently on the metaphase plate.*		
Meiosis I					
Meiosis II					

2. Fill in the chart to summarize the major similarities and differences in the two types of cell division (mitosis vs. meiosis). For similarities, include the event(s) that always happen(s) at that stage, no matter which of the cell division cycles you're describing.

	Interphase	Prophase	Metaphase	Anaphase	Telophase
a. What similarities do you see?					
b. What differences do you see?					

c. If the amount of DNA in a somatic cell equals C during G_1 of interphase, how much DNA is present in the cell during each of the phases of mitosis and meiosis?

Amount of DNA in:	Interphase	Prophase	Metaphase	Anaphase	Telophase
Mitosis					
Meiosis I					
Meiosis II					

3. How do the similarities in prophase of mitosis and meiosis compare to the similarities in telophase of mitosis and meiosis?

4. At what stage(s) does/do most of the differences among mitosis, meiosis I, and meiosis II occur? For what reasons do these differences exist?

Name_____   Course/Section_____

Date_____   Professor/TA_____

Activity 14.1 A Genetics Vocabulary Review

Mendel did not know anything about chromosomes, genes, or DNA. Because modern genetics uses vocabulary that assumes students today understand these ideas, it's helpful to review some key terms.

Match each commonly used genetics term with its appropriate definition or example.

Terms	Definitions and Examples
_____ heterozygous	a. Blue-eyed blonde mates with brown-eyed brunette
_____ homozygous	b. *BB* or *bb*
_____ monohybrid cross	c. not on sex chromosomes
_____ autosomal	d. blue or brown eyes
_____ genotype	e. *Bb*
_____ phenotype	f. locus on a chromosome that codes for a given polypeptide
_____ gene	g. Blonde mates with brunette
_____ allele	h. *BB, Bb,* or *bb*
_____ dihybrid cross	i. Males have only one for each gene on the X chromosome

Name_____   Course/Section _____

Date_____   Professor/TA _____

 Activity 14.2 Modeling meiosis: How can diploid organisms produce haploid gametes?

Integrate your understanding of meiosis (Chapter 13) and of basic Mendelian principles (Chapter 14) to develop a dynamic model of meiosis. When you've completed the model, use it to explain what aspects of meiosis account for Mendel's laws of segregation and independent assortment.

Building the Model

Working in groups of three or four, construct a dynamic (claymation-type) model of meiosis for the organism described on the next page. You may use the materials provided in class or devise your own.

What genetic and chromosomal traits does your organism have?

1. Your individual is male/female (choose one). Females are XX and males are XY. For simplicity, assume that the individual is diploid with $2n = 6$, including the sex chromosomes. On one pair of autosomes (the nonsex chromosomes), the individual is heterozygous for hair color (B = brown and dominant, b = blonde and recessive). On another pair of autosomes, the organism is heterozygous for hair structure (C = curly and dominant, c = straight and recessive). Assume further that the individual's mother was homozygous dominant for both traits and the father was homozygous recessive for both.

   a. Is your individual's hair curly or straight? Brown or blonde?

   b. What did the individual's mother's hair look like? What did the father's hair look like?

   c. What chromosomes and alleles were in the egg and the sperm that gave rise to your individual?

What does the nucleus contain?

To answer this question, develop a model of a cell from your individual.

- Use chalk on a tabletop or a marker on a large sheet of paper to draw a cell's membrane and its nuclear membrane. The nucleus should be at least 9 inches in diameter.
- Use playdough or cutout pieces of paper to represent your individual's chromosomes. Indicate the placement of genes on the chromosomes. Put all the chromosomes from your individual into the nucleus.
- Make a key for your model that indicates how alleles are designated and which of the chromosomes are maternal versus paternal contributions.

Then develop a model of the meiotically active cell.

- Make an identical copy of the original cell. This will be the "active" cell—that is, the one that undergoes meiosis.
- Using the "active" cell only, develop a dynamic model of meiosis. To do this, actively move the chromosomes of this one cell through a complete round of meiosis in a sex cell. (Sex cells are the cells of the body that give rise to gametes: ova or sperm.)
- Use your model to demonstrate meiosis to another student group or to your instructor. Then use your model to answer the questions on the next page.

When developing and explaining your model, be sure to include definitions or descriptions of all these terms and structures:

diploid	dominant allele
2n/n	genotype
chromosome	maternal
chromatid	paternal
chromatin	spindle
centromere (kinetochore)	spindle fibers
autosome	nuclear membrane
sex chromosome	nucleolus
sex cell	phenotype
autosome	heterozygous
crossing over	homozygous
synapsis	law of segregation
recessive allele	law of independent assortment

Name_____   Course/Section_____

What are the products of meiosis?

2. From a single sex cell going through meiosis, how many daughter cells are produced?

3. For your model organism or individual (defined in question 1), how many different kinds of gametes can be produced from a single cell undergoing meiosis? (Assume no crossing over occurs.)

4. Your individual is heterozygous for two genes on separate pairs of homologous chromosomes. His/her genotype is *CcBb*. Given this information alone, how many different kinds of gametes could this individual produce? (Again, assume no crossing over occurs.)

5. Compare your answer to question 4 with your answer to question 3. How do the numbers of different kinds of gametes in your answers compare? Explain any difference.

14.2 Test Your Understanding

What aspect(s) of meiosis account(s) for:

1. Mendel's law of segregation?

2. Mendel's law of independent assortment?

Name_____   Course/Section_____

Date_____   Professor/TA_____

Activity 14.3 A Quick Guide to Solving Genetics Problems

Over the years, rules have been developed for setting up genetics problems and denoting genes and their alleles in these problems. This activity provides a quick review of some of these rules. After you have read through all of this material, complete Activities 14.4, 15.1, and 15.2.

Basic Assumptions to Make When Solving Genetics Problems

1. Are the genes linked?

If the problem does not (a) indicate that the genes are linked or (b) ask whether the genes are (or could be) linked, then you should assume that the genes are not linked.

2. Are the genes sex-linked?

Similarly, if the problem does not (a) indicate that the genes are sex-linked (that is, on the X chromosome) or (b) ask whether the genes are (or could be) on the X chromosome (or Y chromosome), then you should assume that the genes are on autosomes and are not sex-linked.

3. Is there a lethal allele?

If a gene is lethal, then you should assume that the offspring that get the lethal allele (if dominant) or alleles (if homozygous recessive) do not appear; that is, they are not born, do not hatch, and so on. Therefore, they are not counted among the offspring. An obvious exception is lethal genes that have their effect late in life. If this is the case, however, it should be noted in the question.

4. Are the alleles dominant, recessive, or neither?

Unless the problem states otherwise, assume that capital letters (*BB*, for example) designate dominant alleles and lowercase letters (*bb*, for example) indicate recessive alleles. When there is codominance or incomplete dominance, the alleles are usually designated by the same capital letter and each one is given a superscript (for example, $C^R C^W$ in Figure 14.10, page 272, of *Biology,* 8th edition).

5. How are genotypes written?

Assume a gene for fur color in hamsters is located on the number 1 pair of homologous autosomes. Brown fur (B) is dominant over white fur (b). The genotype for fur color can be designated in different ways:

a. The alleles can be shown associated with the number 1 chromosome. In this notation, an individual heterozygous for this gene is designated as $I^B I^b$.

b. Most commonly, this notation is simplified to *Bb*.

In problems that involve sex-linked genes, the chromosomes are always indicated—for example, $X^A X^a$ and $X^a Y$.

6. What information do you need to gather before trying to solve a genetics problem?

Before trying to solve any problem, answer these questions:

a. What information is provided? For example:
    *   What type of cross is it? Is it a monohybrid or dihybrid cross?
    *   Are the genes sex-linked or autosomal?
    *   Linked or unlinked?

b. What does the information provided tell you about the gene(s) in question? For example:
    *   What phenotypes can result?
    *   How many alleles does the gene have?
    *   Are the alleles of the gene dominant? Recessive? Codominant?

c. Does the question supply any information about the individuals' genotypes? If so, what information is provided?
    *   Grandparent information?
    *   Parental (P) information?
    *   Gamete possibilities?
    *   Offspring possibilities?

Solving Genetics Problems

1. What is a Punnett square?

Punnett squares are frequently used in solving genetics problems. A Punnett square is a device that allows you to determine all the possible paired combinations of two sets of characteristics. For example, if you wanted to determine all the possible combinations of red, blue, and green shirts with red, blue, and green pants, you could set up this Punnett square:

		Shirts		
		Red shirt	Blue shirt	Green shirt
Pants	Red pants			
	Blue pants			
	Green pants			

Similarly, if you wanted to determine the probability of a male (XY) and a female (XX) having a son or a daughter, you would first determine the possible gametes each could produce and then set up a Punnett square to look at all the possible combinations of male and female gametes. Here, meiosis dictates that the female's gametes get one of her X chromosomes or the other. In the male, the gametes get either the X chromosome or the Y. As a result, the Punnett square would look like this:

		Female's gamete possibilities	
		X	X
Male's gamete possibilities	X	XX	XX
	Y	XY	XY

2. **If you know the parents' genotypes, how can you determine what types of offspring they will produce?**

   a. **Autosomal genes:** For an autosomal gene that has the alleles *A* and *a*, the three possible genotypes are *AA*, *Aa*, and *aa*.

All possible combinations of matings and offspring for two individuals carrying the autosomal gene with alleles *A* and *a* are shown in the figure below.

If you know how to solve these six crosses you can solve any problem involving one or more autosomal genes.

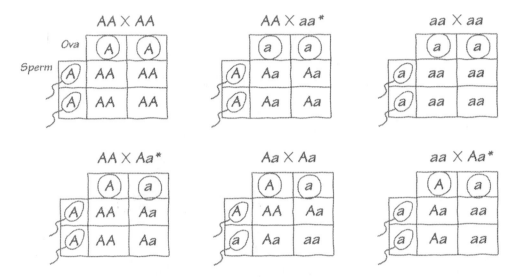

Note: If you take sex into account, there are actually nine possible combinations of matings:

Male genotypes	Female genotypes		
	AA	*Aa*	*aa*
AA	AA x AA	AA x Aa	**AA x aa**
Aa	Aa x AA	Aa x Aa	Aa x aa
aa	**AA x aa**	aa x Aa	aa x aa

Because the results of reciprocal autosomal matings—e.g., *AA* male with *aa* female and *aa* male with *AA* female—are the same, only one of each reciprocal type is included in the six combinations above.

b. **Sex-linked genes:** For sex-linked genes that have two alleles, e.g., $w+$ and w, females have three possible genotypes: $X^{w+}X^{w+}$, $X^{w+}X^{w}$, and $X^{w}X^{w}$. Males have only two possible genotypes: $X^{w+}Y$ and $X^{w}Y$. All the possible combinations of matings and offspring for a sex-linked trait are listed in the next figure. If you know how to solve these six single-gene crosses, then you can solve any genetics problem involving sex-linked genes.

All possible combinations of matings for two individuals with a sex-linked gene are shown in the figure below. Fill in the Punnet squares to determine all possible combinations of offspring.

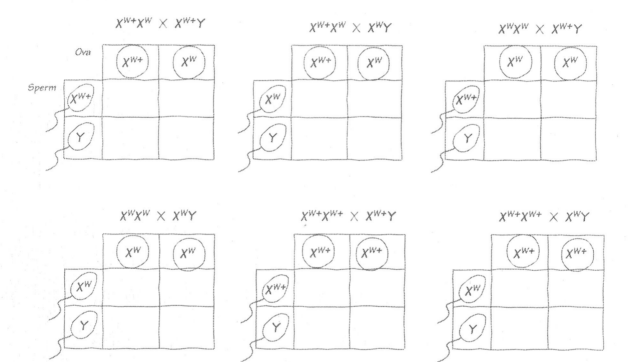

c. **Multiple genes:** Remember, if genes are on separate chromosomes, then they assort independently in meiosis. Therefore, to solve a genetics problem involving multiple genes, where each gene is on a separate pair of homologous chromosomes:

- Solve for each gene separately.
- Determine probabilities for combination (multiple-gene) genotypes by multiplying the probabilities of the individual genotypes.

Example

What is the probability that two individuals of the genotype *AaBb* and *aaBb* will have any *aabb* offspring?

To answer this, solve for each gene separately.

A cross of *Aa* × *aa* could produce the following offspring:

	A	a
a	Aa	aa
a	Aa	aa

¹/2 *Aa* and ¹/2 *aa* offspring

A cross of *Bb* × *Bb* could produce the following offspring:

	B	b
B	BB	Bb
b	Bb	bb

¹/4 *BB*, ¹/2 *Bb*, and ¹/4 *bb* offspring

The probability of having any *aabb* offspring is then the probability of having any *aa* offspring times the probability of having any *bb* offspring.

The probability is ¹/2 × ¹/4 = ¹/8

Name_____   Course/Section_____

Date_____   Professor/TA_____

Activity 14.4 How can you determine all the possible types of gametes?

To solve genetics problems in which genotypes are given, you must first know what types of gametes each organism can produce.

1. How many different kinds of gametes can individuals with each of the following genotypes produce?

   a. *AA*   e. *AaBb*

   b. *aa*   f. *AaBbCC*

   c. *Aa*   g. *AaBbCc*

   d. *AaBB*   h. *AaBbCcDdEeFf*

2. Based on your answers in question 1, propose a general rule for determining the number of different gametes organisms like those described in question 1 can produce.

3. Two individuals have the genotypes *AaBbCcDd*.

    a. How many different types of gametes can each produce?

    b. What are these gametes?

    c. You set up a Punnett square using all the possible gametes for both individuals. How many "offspring squares" are in this Punnett square?

    d. If you completed this Punnett square, how easy would it be to find all the "offspring squares" that contain the genotype *AaBBccDd*?

    e. Given that the genes are all on separate pairs of homologous chromosomes, what other method(s) could you use to determine the probability of these individuals having any offspring with the genotype *AaBbccDd*?

Name_____   Course/Section_____

Date_____   Professor/TA_____

 Activity 15.1 Solving Problems When the Genetics Are Known

Refer to Activity 14.3 and to Chapters 14 and 15 in *Biology*, 8th edition, to complete this activity.

  1. An organism that has the genotype *AaBbCc* is crossed with an organism that has the genotype *AABbCc*. Assume all genes are on separate sets of chromosomes (that is, they are not linked).

  a. What is the probability that any of the offspring will have the genotype *AABBCC*? (*Hint:* To get the answer, consider the six possible types of autosomal crosses. Determine the individual probabilities of getting *AA* offspring from the monohybrid cross. Then do the same to determine the probabilities of getting *BB* offspring and *CC* offspring. Multiply these probabilities together.)

  b. What is the probability that any of the offspring will have the genotype *AaBbcc*?

  2. Consider the cross *AaBbCcddEe* × *AABBccDDEe*.

  a. What is the probability that any offspring will have the genotype *AaBBCcDdEE*?

  b. What is the probability that any offspring will have the genotype *AABBCCDDee*?

3. In fruit flies (*Drosophila melanogaster*), the most common eye color is red. A mutation (or allele) of the gene for eye color produces white eyes. The gene is located on the X chromosome.

   a. What is the probability that a heterozygous red-eyed female fruit fly mated with a white-eyed male will produce any white-eyed offspring?

   b. What is the probability that the mating in part a will produce any white-eyed females?

   c. What is the probability that this mating will produce any white-eyed males?

4. A heterozygous brown-eyed human female who is a carrier of color blindness marries a blue-eyed male who is not color-blind. Color blindness is a sex-linked trait. Assume that eye color is an autosomal trait and that brown is dominant over blue. What is the probability that any of the offspring produced have the following traits?

   a. Brown eyes

   b. Blue eyes

   c. Color blindness

   d. Color-blind males

   e. Brown-eyed, color-blind males

   f. Blue-eyed, color-blind females

   g. What is the probability that any of the males will be color-blind?

   h. Why do males show sex-linked traits more often than females?

Name_____   Course/Section_____

Date_____   Professor/TA_____

Activity 15.2 Solving Problems When the Genetics Are Unknown

An understanding of Mendelian genetics allows us to determine the theoretical probabilities associated with normal transmission of autosomal and sex-linked alleles during reproduction. This understanding provides us with strategies for solving genetics problems. In real-life situations, geneticists use these strategies to determine the genetics behind specific phenotypic traits in organisms. They do this by conducting controlled crosses of experimental organisms (e.g., *Drosophila*) or by analyzing family pedigrees (as for humans).

Controlled Crosses

Two problems are presented below. In each, you are given:

a. "Wild population"—the phenotypic characteristics of a wild population of fruit flies that were trapped randomly on a remote island.

b. "Cross 1, 2, etc."—the phenotypic characteristics of offspring from a controlled cross. The phenotypes of the parents are indicated after each cross—e.g., "Cross 1: Male Ambler × Female Wild Type."

For each of the problems, analyze the results in each cross and answer the questions that follow.

1. **Problem One**

Wild population	Wild type	Ambler	Total
Male	33	17	50
Female	31	19	50
Total	64	36	100

Cross 1: Male Ambler × Female Wild Type

Offspring Vial 1	Wild type	Ambler	Total
Male	29	24	53
Female	29	31	50
Total	58	55	113

a. What does cross 1 tell you about dominance versus recessiveness of the alleles?

b. What does cross 1 tell you about placement of the alleles on autosomes vs. sex chromosomes?

Cross 2: Female Ambler × *Male Wild Type*

Offspring Vial 2	Wild type	Ambler	Total
Male	0	32	32
Female	32	0	32
Total	32	32	64

a. What does cross 2 tell you about dominance versus recessiveness of the alleles?

b. What does cross 2 tell you about placement of the alleles on autosomes vs sex chromosomes? (In your answer show the chromosomal genotypes for the parents in this cross.)

2. **Problem Two**

Mt = Monocle; Bt = Bifocal; Tr =Trifocal; Sp = Spinner; Sh = Shing

Wild Population	Mt, Sp	Mt, Sh	Bt, Sp	Bt, Sh	Tr, Sp	Tr, Sh	Total
Male	10	6	6	0	22	3	47
Female	19	1	9	1	20	4	54
Total	29	7	15	1	42	7	101

Cross 1: Bifocal, Spinner Female × *Monocle, Shiny Male*
Mt = Monocle; Bt = Bifocal; Tr =Trifocal; Sp = Spinner; Sh = Shing

Offspring Vial 1	Mt, Sp	Mt, Sh	Bt, Sp	Bt, Sh	Tr, Sp	Tr, Sh	Total
Male	0	0	0	0	31	34	65
Female	0	0	0	0	34	38	72
Total	0	0	0	0	65	72	137

a. What does cross 1 tell you about dominance versus recessiveness of the alleles?

b. What does cross 1 tell you about placement of the alleles on autosomes vs. sex chromosomes? (In your answer show the chromosomal genotypes for the parents in this cross.)

Name_____   Course/Section_____

Cross 2: Monocle, Spinner Female × Trifocal, Spinner Male
Mt = Monocle; Bt = Bifocal; Tr =Trifocal; Sp = Spinner; Sh = Shing

Offspring Vial 2	Mt, Sp	Mt, Sh	Bt, Sp	Bt, Sh	Tr, Sp	Tr, Sh	Total
Male	8	8	0	0	8	8	32
Female	23	0	0	0	15	0	38
Total	31	8	0	0	23	8	70

a. What does cross 2 tell you about dominance versus recessiveness of the alleles?

b. What does cross 2 tell you about placement of the alleles on autosomes vs. sex chromosomes? (In your answer show the chromosomal genotypes for the parents in this cross.)

Analysis of Pedigrees

Analyze the pedigree and answer the questions that follow.

The diagram below shows a pedigree of three generations in a family. Black circles/squares indicate persons with a genetic disorder. A square indicates a male and a circle indicates a female. The two males in generation 1 are siblings.

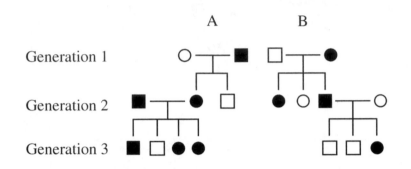

3. Looking only at the generation 2 offspring (of the two generation 1 brothers), what can you say about the gene(s) controlling the genetic disorder? Is the disorder caused by a gene that is dominant or recessive, autosomal or sex-linked?

4. What additional information do you gain from examining the generation 3 offspring?

Name_____   Course/Section_____

Date_____   Professor/TA_____

Activity 15.3 How can the mode of inheritance be determined experimentally?

Outline the experimental crosses you would need to make to solve each problem.

1. Three new traits have been discovered in a population of *Drosophila*:

    - Tapping (a behavioral mutant in which the fly taps one foot constantly)
    - Single stripe (a pigmentation change that leads to a long stripe down the fly's back)
    - Angular (causes angular bends in bristles that are normally straight)

The positions of the three genes on the chromosomes are unknown. Given two pure breeding (homozygous) lines and using an initial cross of normal, normal, normal females with tapping, single stripe, angular males, describe the appropriate genetic experiments needed to establish whether any of these traits are caused by genes that are:

   a. Autosomal or sex-linked

   b. Linked on the same chromosome or unlinked

2. A genetics student chose a special project involving a three-gene cross to check the relative positions and map distances separating three genes in *Drosophila* that she thought were all on the third chromosome. To do this, she mated *Drosophila* females that were homozygous for the recessive genes *cu* (curled), *sr* (striped), and *e* (ebony) with males that were homozygous for the wild type, cu^+ (straight), sr^+ (not striped), and e^+ (gray). She then mated (testcrossed) the F_1 females with homozygous recessive curled, striped, ebony males.

Here are the phenotypic results of the testcross:

straight, gray, not striped	820
curled, ebony, striped	810
straight, ebony, striped	100
curled, gray, not striped	97
straight, ebony, not striped	80
curled, gray, striped	90
straight, gray, striped	1
curled, ebony, not striped	2
Total	2,000

a. How are the three genes arranged on the chromosomes?

b. What evidence allows you to answer the question in part a?

c. If any of the genes are linked, how far apart are they on the chromosome? How can you determine this?

Name_____   Course/Section_____

Date_____   Professor/TA_____

Activity 16.1 Is the hereditary material DNA or protein?

Accumulating and Analyzing the Evidence

Build a concept map to review the evidence used to determine that DNA was the genetic material, the structure of DNA, and its mode of replication. Keep in mind that there are many ways to construct a concept map.

- First, develop a separate concept map for each set of terms (A to D on the next page). Begin by writing each term on a separate sticky note or sheet of paper.
- Then organize each set of terms into a map that indicates how the terms are associated or related.
- Draw lines between the terms and add action phrases to the lines to indicate how the terms are related.

Here is an example:

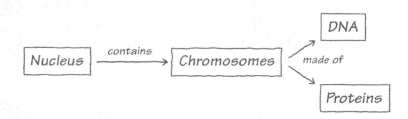

- After you have completed each of the individual concept maps, merge or interrelate the maps to show the overall logic used to conclude that DNA (not protein) is the hereditary material.
- When you have completed the overall concept map, answer the questions.

Terms

Map A
Griffith

mice

S strain of *Streptococcus*

R strain of *Streptococcus*

live

heat-killed

transformation

Avery, McCarty, and MacLeod

DNA

protein

Map B
Hershey and Chase

bacteria

bacteriophage (phage) (only a protein
   and DNA)

^{35}S

^{32}P

Waring blender

high-velocity centrifugation

Map C
Watson and Crick

X-ray crystallography

Chargaff's rule

purine structure

pyrimidine structure

H bonds

phosphate sugar backbone

Map D
Meselson and Stahl

conservative

dispersive

semiconservative

nucleic acid bases

^{14}N

^{15}N

bacteria

density equilibrium centrifugation

replication

Name_____   Course/Section_____

1. In the early to mid-1900s, there was considerable debate about whether protein or DNA was the hereditary material.

   a. For what reasons did many researchers assume that protein was the genetic material?

b. What key sets of experiments led to the understanding that, in fact, DNA and not protein was the hereditary material?	c. What evidence did each experiment provide?

2. Watson and Crick were the first to correctly describe the structure of DNA. What evidence did they use to do this? How did they use this evidence to put together or propose the structure of DNA?

3. How did the results of Meselson and Stahl's experiments show that DNA replicates semiconservatively? To answer this, answer the following questions.

    a. Diagram the results that would be expected for each type of replication proposed.

    b. What evidence allowed Meselson and Stahl to eliminate the conservative model?

    c. What evidence allowed them to eliminate the dispersive model?

Name_____   Course/Section_____

16.1 Test Your Understanding

An *E. coli* cell that contains a single circular chromosome made of double-stranded DNA is allowed to replicate for many generations in ^{15}N medium until all of the *E. coli* cells' DNA is labeled with ^{15}N. One *E. coli* cell is removed from the ^{15}N medium and is placed into medium in which all of the available nucleotides are ^{14}N labeled. The *E. coli* cell is allowed to replicate until eight *E. coli* are formed.

1. Given this situation, which of the following is true?
    a. Some ^{15}N-labeled DNA will be found in all eight cells.
    b. Some ^{15}N-labeled DNA will be found in only four of the cells.
    c. Some ^{15}N-labeled DNA will be found in only two of the cells.
    d. Some ^{15}N-labeled DNA will be found in only one of the cells.

2. To explain your answer, draw the sequence of events that occurred.

Name_____   Course/Section_____

Date_____   Professor/TA_____

Activity 16.2 How does DNA replicate?

Working in groups of three or four, construct a dynamic (working or active) model of DNA replication. You may use the materials provided in class or devise your own.

Building the Model

- Develop a model of a short segment of double-stranded DNA.
- Include a key for your model that indicates what each component represents in the DNA molecule—for example, adenine, phosphate group, deoxyribose.
- Create a dynamic (claymation-type) model of replication. Actively move the required bases, enzymes, and other components as needed to model replication of your DNA segment.

Your model should describe the roles and relationships of all the following enzymes and structures in replication:

parental DNA

nucleotide excision repair

daughter DNA

mutation

antiparallel strands

single-stranded DNA-binding proteins

leading strand

telomeres

lagging strand

telomerase

$5'$ end

$3'$ end

$3' \rightarrow 5'$ versus $5' \rightarrow 3'$

nitrogenous bases A, T, G, C

replication fork

replication bubble

Okazaki fragment

DNA polymerase

helicase

DNA ligase

primase

RNA primers

origin of replication

Use your model to answer the questions.

1. Explain how Meselson and Stahl's experiments support the idea that DNA replication is semiconservative.

2. A new form of DNA is discovered that appears to be able to replicate itself both in the 3′ → 5′ direction and in the 5′ → 3′ direction. If this is true, how would this newly discovered DNA replication differ from DNA replication as we know it?

3. Amazingly, an alien species of cellular organism is found alive in the remains of a meteorite that landed in the Mojave Desert. As a scientist, you are trying to determine whether this alien life-form uses DNA, protein, or some other type of compound as its hereditary material.

   a. What kinds of experiments would you propose to determine what the hereditary material is?

   b. Assuming that the hereditary material turns out to be similar to our DNA, describe the simplest experiments you could run to try to determine if it is double-stranded like our DNA, triple-stranded, or something else.

4. Some researchers estimate that the mutation rate for any given gene (or its DNA) in certain strains of bacteria is about 10^{-8}. This means that one error or mutation in a given gene is introduced for every 100-million cell divisions.

   a. What can cause mistakes in replication?

   b. How are such mistakes normally corrected?

Name_____   Course/Section_____

Date_____   Professor/TA_____

Activity 17.1 Modeling transcription and translation: What processes produce RNA from DNA and protein from mRNA?

Create a model of the processes of transcription and translation. Your model should be a dynamic (working or active) representation of the events that occur first in transcription in the nucleus and then in translation in the cytoplasm.

For the purposes of this activity, assume there are *no introns* in the mRNA transcript.

When developing and explaining your model, be sure to include definitions or descriptions of the following terms and structures:

gene

DNA

nucleotides: A, T, G, and C versus A, U, G, and C

RNA modification(s) after transcription

mRNA

RNA polymerase

poly(A) tail

5′ cap

translation

protein synthesis

ribosome (large versus small subunit)

A, P, and E sites

tRNA

rRNA

start codon (methionine)

aminoacyl-tRNA synthetase

amino acids (see Figure 17.4, page 329, in *Biology*, 8th edition)

peptidyl transferase

polypeptide

codons

stop codons

anticodons

initiation

elongation

termination

polypeptide

Building the Model

- Use chalk on a tabletop or a marker on a large sheet of paper to draw a cell's plasma membrane and nuclear membrane. The nucleus should have a diameter of about 12 inches.

- Draw a DNA molecule in the nucleus that contains the following DNA sequence:

  Template strand   3′ TAC TTT AAA GCG ATT 5′

  Nontemplate strand   5′ ATG AAA TTT CGC TAA 3′

- Use playdough or cutout pieces of paper to represent the various enzymes, ribosome subunits, amino acids, and other components.

- Use the pieces you assembled to build a dynamic (claymation-type) model of the processes of transcription and translation.

- When you feel you have developed a good working model, use it to explain the processes of transcription and translation to another student or to your instructor.

Use your model of transcription and translation to answer the questions.

1. How would you need to modify your model to include intron removal? Your explanation should contain definitions or descriptions of the following terms and structures:

   pre-mRNA   exons

   RNA splicing   spliceosome

   introns

2. If 20% of the DNA in a guinea pig cell is adenine, what percentage is cytosine? Explain your answer.

Activity 17.1

Name_____   Course/Section_____

3. A number of different types of RNA exist in prokaryotic and eukaryotic cells. List the three main types of RNA involved in transcription and translation. Answer the questions to complete the chart.

a. Type of RNA	b. Where is it produced?	c. Where and how does it function in cells?

4. Given your understanding of transcription and translation, fill in the blanks below and indicate the 5' and 3' ends of each nucleotide sequence. Again, assume no RNA processing occurs.

Nontemplate strand of DNA: 5′ A T G T A T G C C A A T G C A 3′

Template strand of DNA: __′ T _ _ _ _ _ _ _ _ _ _ _ _ _ _ __′

mRNA: __′ A _ _ _ _ U _ _ _ _ _ _ _ _ _′

Anticodons on complementary tRNA: __′ _ _ _ / _ _ _ / _ _ _ / _ _ _ / _ _ _ / _ _′

5. Scientists struggled to understand how four bases could code for 20 different amino acids. If one base coded for one amino acid, the cell could produce only four different kinds of amino acids (4^1). If two bases coded for each amino acid, there would be four possible choices (of nucleotides) for the first base and four possible choices for the second base. This would produce 4^2 or 16 possible amino acids.

a. What is the maximum number of *three-letter codons* that can be produced using only four different nucleotide bases in DNA?

b. How many different codons could be produced if the codons were four bases long?

Mathematical logic indicates that at least three bases must code for each amino acid. This led scientists to ask:

- How can we determine whether this is true?
- Which combinations of bases code for each of the amino acids?

To answer these questions, scientists manufactured different artificial mRNA strands. When placed in appropriate conditions, the strands could be used to produce polypeptides.

Assume a scientist makes three artificial mRNA strands:

(*x*) 5' AAAAAAAAAAAAAAAAAAAAAAAAAAA 3'

(*y*) 5' AAACCCAAACCCAAACCCAAACCCAAA 3'

(*z*) 5' AUAUAUAUAUAUAUAUAUAUAUAUAU 3'

When he analyzes the polypeptides produced, he finds that:

x produces a polypeptide composed entirely of lysine.

y produces a polypeptide that is 50% phenylalanine and 50% proline.

z produces a polypeptide that is 50% isoleucine and 50% tyrosine.

c. Do these results support the three-bases-per-codon or the four-bases-per-codon hypothesis? Explain.

d. The type of experiment just described was used to discover the mRNA nucleotide codons for each of the 20 amino acids. If you were doing these experiments, what sequences would you try next? Explain your logic.

6. Now that the complete genetic code has been determined, you can use the strand of DNA shown here and the codon chart in Figure 17.4 (page 329) in *Biology*, 8th edition to answer the next questions.

Original template strand of DNA: 3' TAC GCA AGC AAT ACC GAC GAA 5'

a. If this DNA strand produces an mRNA, how does the sequence of the mRNA read from 5' to 3'?

b. For what sequence of amino acids does this mRNA code? (Assume it does not contain introns.)

c. The chart below lists five point mutations that may occur in the original strand of DNA. What happens to the amino acid sequence or protein produced as a result of each mutation? (*Note:* Position 1 refers to the first base at the 3′ end of the transcribed strand. The last base in the DNA strand, at the 5′ end, is at position 21.)

Original template strand: 3′ TAC GCA AGC AAT ACC GAC GAA 5′

Mutation	Effect on amino acid sequence
i. Substitution of T for G at position 8.	
ii. Addition of T between positions 8 and 9.	
iii. Deletion of C at position 15.	
iv. Substitution of T for C at position 18.	
v. Deletion of C at position 18.	
vi. Which of the mutations produces the greatest change in the amino acid sequence of the polypeptide coded for by this 21-base-pair gene?	

7. Sickle-cell disease is caused by a single base substitution in the gene for the beta subunit of hemoglobin. This base substitution changes one of the amino acids in the hemoglobin molecule from glutamic acid to valine. Look up the structures of glutamic acid (glu) and valine (val) on page 79 of *Biology*, 8th edition. What kinds of changes in protein structure might result from this substitution? Explain.

8. Why do dentists and physicians cover patients with lead aprons when they take mouth or other X-rays?

17.1 Test Your Understanding

During DNA replication, which of the following would you expect to be true? Explain your answers.

T/F   1.   More ligase would be associated with the lagging strand than with the leading strand.

T/F   2.   More primase would be associated with the lagging strand than with the leading strand.

T/F   3.   More helicase would be associated with the lagging strand than with the leading strand.

T/F   4.   DNA ligase links the 3′ end of one Okazaki fragment to the 5′ end of another Okazaki fragment in the lagging strand.

T/F   5.   In the lagging strand, the enzyme DNA polymerase III that produces the next Okazaki fragment also removes the short segment of primer RNA on the previous Okazaki fragment.

6.   You obtain a sample of double-stranded DNA and transcribe mRNA from this DNA. You then analyze the base composition of each of the two DNA strands and the one mRNA strand, and get the following results. The numbers indicate the percentage of each base in the strand:

	A	G	C	T	U
strand 1	40.1	28.9	9.9	0.0	21.1
strand 2	21.5	9.5	29.9	39.1	0.0
strand 3	40.0	29.0	9.7	21.3	0.0

a.   Which of these strands must be the mRNA? Explain.

b.   Which one is the template strand for the mRNA? Explain.

7. In a new experiment, you obtain a different sample of double-stranded DNA and transcribe mRNA from this DNA. You then analyze the base composition of each of the two DNA strands and the one mRNA strand, and get the following results. The numbers indicate the percentage of each base in the strand:

	A	G	C	T	U
strand 1	29.1	39.9	31.0	0.0	0.0
strand 2	0.0	30.0	39.8	30.2	0.0
strand 3	29.4	39.4	31.2	0.0	0.0

a. Which of these strands could be the mRNA? Explain.

b. Which one must be the template strand for the mRNA? Explain.

8. Cystic fibrosis transmembrane conductance regulator (CFTR) proteins function in cell membranes to allow chloride ions to cross cell membranes. Individuals with cystic fibrosis (CF) have abnormal CFTR proteins that do not allow Cl⁻ to move across cell membranes. Chloride channels in the membrane are essential to maintain osmotic balance inside cells. Without properly functioning Cl⁻ channels, water builds up inside the cell. One result is a thickening of mucous in lungs and air passages.

You are doing research on a different disease, and you hypothesize that it may also be due to a defect in an ion channel in the cell membrane.

a. Diagram or model production of a normal membrane ion channel.

b. Based on your understanding of cell membrane structure and function, propose at least three different alterations that could result in a nonfunctional or missing ion channel.

c. What questions would you need to answer to determine which of these alterations may be correct?

Name_____   Course/Section_____

Date_____   Professor/TA_____

Activity 18.1 How is gene expression controlled in bacteria?

Fill in the chart to organize what we know about the *lac* and *trp* operons.

Operon:	*lac*		*trp*	
Is the metabolic pathway anabolic or catabolic?	*Catabolic* *Breaks down lactose*		*Anabolic* *Synthesizes tryptophan*	
What regulatory genes are associated with the operon, and what functions does each serve?	Genes:	Functions:	Genes:	Functions:
What structural genes are included in each operon, and what does each produce?	Genes:	Products:	Genes:	Products:
Is the operon inducible or repressible?				
Is the repressor protein produced in active or inactive form?				
The repressor protein becomes active when it interacts with:				

Name_____   Course/Section_____

Date_____   Professor/TA_____

 Activity 18.2 Modeling the *lac* and *trp* operon systems: How can gene expression be controlled in prokaryotes?

Using the information in Activity 18.1 and in Chapter 18 of *Biology*, 8th edition, construct a model or diagram of the normal operation of both the *lac* and *trp* operon systems.

In your models or diagrams, be sure to include these considerations:

      regulatory and structural genes

      inducible versus repressible control

      anabolic versus catabolic enzyme activity

      negative versus positive controls

Use your model to answer the questions.

1. Under what circumstances would the *lac* operon be "on" versus "off"? The *trp* operon?

2. How are the *lac* and *trp* operons similar (in structure, function, or both)?

3. What are the key differences between the *lac* and *trp* operons?

4. What advantages are gained by having genes organized into operons?

5. Strain X of *E. coli* contains a mutated *lac* regulatory gene on its bacterial genome. As a result, the gene produces a nonfunctional *lac* repressor protein. You add a plasmid (an extra circular piece of double-stranded DNA) to these cells. The plasmid contains a normal regulatory gene and a normal *lac* operon.

Build a model or diagram of what one of these modified *E. coli* cells would look like. Then answer the questions and use your model or diagram to explain your answers.

a. Before the addition of the plasmid, would the *E. coli* strain X cells be able to produce the enzymes for lactose digestion? Explain.

b. After the addition of the plasmid, would the plasmid's *lac* operon produce the enzymes for lactose digestion constitutively (all the time) or only when lactose was the available sugar source? Explain.

c. After the addition of the plasmid, would the bacterial genome's *lac* operon produce the enzymes for lactose digestion constitutively or only when lactose was the available energy source? Explain.

d. If equal amounts of lactose and glucose were present in the cell, would the *lac* operon in the bacterial DNA be off or on? Would the *lac* operon on the introduced plasmid be off or on? Explain.

Name_____   Course/Section_____

Date_____   Professor/TA_____

Activity 18.3 How is gene activity controlled in eukaryotes?

Human genes cannot all be active at the same time. If they were, all the cells in our bodies would look the same and have the same function(s). For specialization to occur, some genes or gene products must be active while others are turned off or inactive.

1. In eukaryotes, gene expression or gene product expression can be controlled at several different levels. Indicate what types of control might occur at each level of gene or gene product expression.

Level	Types of control
a. The gene or DNA itself	
b. The mRNA product of the gene	
c. The protein product of the mRNA	

18.3 Test Your Understanding

Single-celled organisms such as *Amoeba* and *Paramecia* often live in environments that change quickly. Which of the following types of control allow organisms like *Amoeba* to respond most quickly to frequent short-term environmental changes? Explain your reasoning.

a. Control of mRNA transcription from DNA
b. Control of enzyme concentration by control of the rate of mRNA translation
c. Control of the activity of existing enzymes
d. Control of the amount of DNA present in the cell

Name_____   Course/Section_____

Date_____   Professor/TA_____

Activity 18.4 What controls the cell cycle?

1. Checkpoints in the normal cell cycle prevent cells from going through division if problems occur—for example, if the DNA is damaged.

    a. What forms do the checkpoints take? That is, how do they control whether or not cell division occurs?

    b. In the space below, develop a handout or diagram to explain how these checkpoints work under normal conditions. Your diagram should include a description of each checkpoint, where it acts in the cell cycle, and what each does to control cell division.

    c. Cancer results from uncontrolled cell division. Explain how mutations in one or more of the checkpoints might lead to cancer.

Name_____   Course/Section_____

Date_____   Professor/TA_____

 Activity 19.1 How do viruses, viroids, and prions affect host cells?

1. By definition, viruses are obligate intracellular parasites. What does this mean?

2. In general, how are viruses classified?

3. What is reverse transcriptase?

    a. Where (biologically) was reverse transcriptase first found?

    b. Of what use is reverse transcriptase to viruses? Of what use is it to scientists?

4. Why must viruses invade other cells to reproduce?

    a. Describe the general process. Include a discussion of lytic and lysogenic viral cycles.

b. Which types of viruses are more likely to have a lysogenic phase?

5. More than 100 different viruses can cause the "common cold" in humans. Many of these are rhinoviruses. Other viruses—influenza viruses—cause the flu. While there are many different antibiotics for treating bacterial infections, there are relatively few drugs available to treat viral infections. Explain.

6. a. How do viruses, viroids, and prions differ in terms of both composition and function?

b. Do these differences make it easier or harder to treat viroid and prion infections compared to viral infections? Explain your reasoning.

Name_____   Course/Section_____

Date_____   Professor/TA_____

Activity 20.1 How and why are genes cloned into recombinant DNA vectors?

Develop a model to explain how a human gene can be cloned into a bacterial plasmid. Your model should be a dynamic (working or active) representation of the events that need to occur in order to

- clone the insulin gene into a bacterial plasmid, and
- transform the plasmid into *E. coli.*

When you develop and explain your model, be sure to include definitions or descriptions of the following terms and components:

restriction enzyme(s)   transformation

ligase   *E. coli*

plasmid DNA   marker genes (an antibiotic resistance gene)

human mRNA (the mRNA for insulin)   cloning vector

reverse transcriptase

Building the Model

- Use chalk on a tabletop or a marker on a large sheet of paper to draw at least two test tubes and an *E. coli* cell's plasma membrane. The *E. coli* should have a diameter of at least 12 inches. The test tubes should have a width of at least 6 inches. Use the test tubes for producing the insulin gene and for cloning the gene into the plasmid. Then transform your recombinant plasmid into the *E. coli* cell.
- Use playdough or cutout pieces of paper to represent the enzymes, RNA molecules, and other components.
- Use the pieces you assembled to develop a dynamic (claymation-type) model to demonstrate how a gene can be cloned into a plasmid and how the plasmid can then be transformed into a bacterial cell.
- When you feel you have developed a good working model, demonstrate it to another student or to your instructor.

Use your model to answer the questions.

1. Prior to recombinant gene technology, the insulin required to treat diabetes was obtained from the pancreases of slaughtered farm animals. Because the insulin was from other species, some humans developed immune responses or allergic reactions to it. As recombinant gene technology advanced, researchers explored the possibility of incorporating the human insulin gene into a plasmid that could be transformed into *E. coli*. If this technology was successful, the *E. coli* would produce human insulin that could be harvested from the bacterial culture medium.

   Researchers first needed to isolate the gene for insulin. To do this, they isolated mRNA (rather than DNA) from the beta cells of human pancreas tissue. Using reverse transcriptase, they made double-stranded DNA molecules that were complementary to the mRNA molecules they extracted from the pancreas cells.

   a. Based on what you know about eukaryotic chromosomes and genes, why did researchers choose to isolate mRNA rather than DNA?

   b. What further adjustments might researchers need to make in the DNA molecules produced by reverse transcriptase before the molecules could be incorporated into bacterial plasmids?

c. Not all the DNA molecules produced by reverse transcription from pancreatic mRNA contained the gene for insulin. Some contained other genes. What mechanisms can be used to locate those bacterial colonies that picked up plasmids containing any of the genes produced by reverse transcription from pancreatic mRNAs?

d. What mechanisms can be used to locate bacterial colonies that picked up only plasmids containing the insulin gene?

(*Note:* Today almost all the insulin used for the treatment of human diabetes is produced using recombinant technology.)

Name_____ Course/Section_____

Date_____ Professor/TA_____

Activity 20.2 How can PCR be used to amplify specific genes?

1. Assume you are using PCR (polymerase chain reaction) to make multiple copies of a gene (shaded in grey below).

   DNA containing gene of interest:
   3′ TATAAGACTTACAAATTTGTCCCCATTTTGC5′
   5′ ATATTTCTGAATGTTTAAACAGGGGTAAAACG3′

   On separate sheets of paper, describe the overall process and diagram the results you would obtain for 1, 2, and 3 rounds of PCR replication using the primers ATGTT and CCATT.

   (*Note*: For simplicity we are showing DNA primers that are only 5 bases in length. In actual practice, the DNA primers used are at least 17 bases long. This length is used to help reduce the risk that the primer anneals with [base pairs with] anything other than the specific segment of DNA to be amplified.)

2. PCR is often used in forensics to amplify small amounts of DNA found at crime scenes. The amplified DNA is then tested for differences in RFLP (restriction fragment length polymorphisms) or STR (single tandem repeat) lengths.

   a. Explain what RFLPs and STRs are.

   b. How do STRs compare for unrelated individuals versus for closely related individuals (for example, parent and child or brother and sister)?

   c. How reliable are these types of DNA fingerprinting for identifying individuals? What factors affect their reliability?

1. Which of the following sequences (on one strand of a double-stranded DNA molecule) is likely to be a cleavage site for a restriction enzyme? Explain your answer.
    a. CGTACC
    b. ATGTCG
    c. GATATG
    d. TGCGCA

2. After undergoing electrophoresis, the gel in the figure below shows the RFLP analysis of DNA samples obtained from a crime scene. Bloodstains on a suspect's shirt (B) were analyzed and compared with blood from the victim (V) and from the suspect (S). Are the bloodstains on the shirt from the victim or from the suspect? Explain.

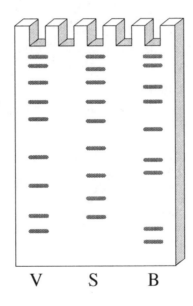

V   S   B

Name_____   Course/Section_____

Date_____   Professor/TA_____

 Activity 21.1 How can we discover the sequence of an organism's DNA?

Bacterial genomes have between 1 million and 6 million base pairs (Mb). Most plants and animals have about 100 Mb; humans have approximately 2,900 Mb. Individual chromosomes may thus contain millions of base pairs. It is difficult to work with DNA sequences this large, so for study purposes the DNA is broken into smaller pieces (approximately 500 to 1,000 bp each). These pieces are sequenced and then the sequenced pieces are examined and aligned based on overlapping sequence homology at their ends.

By comparing the DNA sequences among organisms, scientists can determine:

- what parts of the genomes are most similar among organisms and are therefore likely to have evolved earliest,
- what key differences exist in the genomes that may account for variations among related species, and
- what differences within species exist that may account for development of specific types of disease.

The following activity has been designed to help you understand how genomes are sequenced and how we might use the sequence information.

1. In 1980, Frederick Sanger was awarded the Nobel Prize for inventing the dideoxy method (or Sanger method) of DNA sequencing. A double-stranded DNA segment approximately 700 bp in length is heated (or treated chemically) to separate the two strands. The single-stranded DNA that results is placed into a test tube that contains a 9-to-1 ratio of normal deoxynucleotides to dideoxynucleotides. A dideoxynucleotide has no –OH group at either the 2′ or 3′ carbon. As a result, whenever any dideoxynucleotide (abbreviated ddNTP) is added to the growing DNA strand, synthesis stops at that point. If the ratio of normal to dideoxynucleotides is high enough, where the dideoxynucleotide (rather than the normal deoxynucleotide) will be included in the sequence is random.

    You set up each of four test tubes as noted below:

Tube number	Deoxynucleotide	Dideoxynucleotides
1	dATP, dTTP, dGTP, dCTP	ddATP
2	dATP, dTTP, dGTP, dCTP	ddTTP
3	dATP, dTTP, dGTP, dCTP	ddGTP
4	dATP, dTTP, dGTP, dCTP	ddCTP

Activity 21.1   119

All tubes contain the same single-stranded DNA molecules and the same primers. All other components required for DNA replication, such as enzymes, are present in each tube. You allow the replication to continue for the same length of time in each tube. At the end of the time period, you extract the DNA from each tube and run it on an agarose gel. You dye the gel with ethidium bromide and observe the following banding patterns on the gel. (Note: For this demonstration, we are using a DNA strand that is only 20 bases in length.)

ddATP tube	ddTTP tube	ddGTP tube	ddCTP tube
	—		
	—		
—			
		—	
			—
		—	
—			
	—		
		—	
			—
	—		
—			
—			
	—		
	—		
			—
		—	
—			
			—

a. Which band in the gel contains the shortest DNA strand? What is the identity of its terminal ddNTP?

b. Which band contains the next shortest DNA strand? What is the identity of its terminal ddNTP?

c. Continue reading the terminal ddNTP of each band from shortest to longest to determine the linear sequence of nucleotides in the DNA strand complement. What is the sequence?

2. The Sanger method has been modified so that each ddNTP used is now tagged with an identifying fluorescent marker.

Assume that you run the same experiment that you did earlier. However, this time you combine all of the different nucleotides (both dNTPs and ddNTPs) in the same test tube. You run the products of the reaction on an agarose gel. Indicate the bands you would see on the gel below, using the appropriate colors: ddTTP fluoresces red; ddGTP, yellow; ddCTP, blue; and ddATP, green.

Band sequence for combined experiment

3. To help determine evolutionary relationships among different groups of organisms, scientists compare gene sequences of highly conserved genes. What are "highly conserved genes?" Give examples and indicate what is "highly conserved" and why.

4. What types of DNA do scientists use to determine individual identities of organisms within the same species? Why do they use this type of DNA?

Name_____   Course/Section_____

Date_____   Professor/TA_____

Activity 22.1 How did Darwin view evolution via natural selection?

Darwin is remembered not because he was the first to propose that evolution occurs. Many others had presented this idea before. Instead, he is remembered for defining the mechanism behind evolution—that is, the theory of natural selection. To do this, Darwin integrated, or put together, information from a wide range of sources. Some of this information was provided by others; some he observed on his own.

Working alone or in groups of three or four, construct a concept map of Darwin's view of evolution via natural selection. Be sure to include definitions or descriptions of all the terms in the list below. Keep in mind that there are many ways to construct a concept map.

- Begin by writing each term on a separate sticky note or piece of paper.
- Then organize the terms into a map that indicates how the terms are associated or related.
- Draw lines between terms and add action phrases to the lines that indicate how the terms are related.
- When you finish your map, explain it to another group of students.

Here is an example:

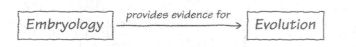

Terms

fact	Darwin	fit individuals
biogeography	vertebrate limb structure	fossil record
gradualism	species population	embryology
uniformitarianism	individual	taxonomy
theory	variability	selective (domestic)
Galápagos Islands	paleontology	breeding
evolution	Malthus	limited resources
homology	population size	struggle for existence
natural selection	environment or	reproduction
analogy	resources	extinction

Use the understanding you gained from creating the concept map to answer the questions.

1. In the 1860s, what types of evidence were available to indicate that evolution had occurred on Earth?

2. How did knowledge of mechanisms of artificial selection (used in developing various strains of domesticated animals and plants) help Darwin understand how evolution could occur?

3. Based on his studies, Darwin made a number of observations; they are listed in the chart. Complete the chart by answering how Darwin made the observations.

Observation	How did Darwin make this observation? That is, what did he read or observe that gave him this understanding?
a. All species populations have the reproduction potential to increase exponentially over time.	
b. The number of individuals in natural populations tends to remain stable over time.	
c. Environmental resources are limited.	
d. Individuals in a population vary in their characteristics.	
e. Much of this variation in characteristics is heritable.	

4. Based on these observations, Darwin made a number of inferences. Which of the observation(s) in question 3 allowed Darwin to make each inference?

Inference	Observations that led to the inference
a. Production of more individuals than the environment can support leads to a struggle for existence such that only a fraction of the offspring survive each generation.	
b. Survival for existence is not random. Those individuals whose inherited traits best fit them to the environment are likely to leave more offspring than less fit individuals.	
c. The unequal ability of individuals to survive and reproduce leads to a gradual change in the population, with favorable characteristics accumulating over the generations.	

5. Based on these observations and inferences, how did Darwin define fitness?

6. How did Darwin define evolution?

7. What is the unit of natural selection—that is, what is selected? What is the unit of evolution—that is, what evolves?

8. In a population of mice, some individuals have brown fur and some have black fur. At present, both phenotypes are equally fit. What could happen to change the relative fitness of the two phenotypes in the population? For example, what could cause individuals with brown fur to show reduced fitness relative to individuals with black fur?

9. Assume you discover a new world on another planet that is full of organisms.

    a. What characteristics would you look for to determine that these organisms arose as a result of evolutionary processes?

    b. What characteristics would you look for to determine that these organisms did *not* arise as the result of evolutionary processes?

10. Why is it incorrect to say: Vertebrates evolved eyes in order to see?

Name_____   Course/Section_____

Date_____   Professor/TA_____

 Activity 22.2 How do Darwin's and Lamarck's ideas about evolution differ?

Early in the 1800s Lamarck proposed a theory of evolution. He suggested that traits acquired during an organism's life—for example, larger muscles—could be passed on to its offspring. The idea of inheritance of acquired characteristics was popular for many years. No such mechanism is implied in Darwin's theory of evolution via natural selection, however. After Darwin published his work, scientists conducted many experiments to disprove the inheritance of acquired traits. By the middle of the 20th century, enough data had accumulated to make even its most adamant supporters give up the idea of inheritance of acquired characteristics.

Given your understanding of both Lamarck's and Darwin's ideas about evolution, determine whether the statements on the next page are more Lamarckian or more Darwinian. If the statement is Lamarckian, change it to make it Darwinian. Here are two example statements and answers.

Examples

A. The widespread use of DDT in the mid-1900s put pressure on insect populations to evolve resistance to DDT. As a result, large populations of insects today are resistant to DDT.

Answer: This is a Lamarckian statement. DDT worked only against insects that had no DDT-resistance genes. The genes for DDT resistance had to be present for insects to survive DDT use in the first place.

Suggested change: Wide-scale use of DDT in the mid-1900s selected against insects that had no resistance to DDT. Only the insects that were resistant to DDT survived. These insects mated and passed their resistance genes on to their offspring. As a result, large populations of insects today are resistant to DDT.

B. According to one theory, the dinosaurs became extinct because they couldn't evolve fast enough to deal with climatic changes that affected their food and water supplies.

Answer: This is a quasi-Lamarckian statement. Organisms do not purposefully evolve. (Genetic recombination experiments are perhaps an exception.) Once you are conceived, your genes are not going to change; that is, you are not going to evolve. The genetic composition of a species population can change over time as certain genotypes are selected against. Genes determine phenotypes. The environmental conditions may favor the phenotype produced by one genotype more than that produced by another.

Suggested change: According to one theory, the dinosaurs became extinct because their physiological and behavioral characteristics were too specialized to allow them to survive the rapid changes in climate that occurred. The climatic changes caused changes in the dinosaurs' food and water supplies. Because none of the dinosaurs survived, the genes and associated phenotypes that would have led to their survival must not have been present in the populations.

Statements

1. Many of the bacterial strains that infect humans today are resistant to a wide range of antibiotics. These resistant strains were not so numerous or common prior to the use of antibiotics. These strains must have appeared or evolved in response to the use of the antibiotics.

2. Life arose in the aquatic environment and later invaded land. Once animals came onto land, they had to evolve effective methods of support against gravity and locomotion in order to survive.

3. A given phenotypic trait—for example, height, speed, tooth structure—(and therefore the genes that determine it) may have positive survival or selective value, negative survival or selective value, or neutral (neither positive nor negative) survival or selective value. Which of these it has depends on the environmental conditions the organism encounters.

4. The children of bodybuilders tend to be much more athletic, on average, than other children because the characteristics and abilities gained by their parents have been passed on to the children.

Activity 22.3 How would you evaluate these explanations of Darwin's ideas?

Unfortunately, even today some people get or give the impression that acquired characteristics can be inherited. As a result, we need to be very careful about how we state our understanding of evolution and evolutionary theory.

To test understanding of Darwin's ideas, this question was included on an exam.

> *4-point question:*
> *In two or three sentences describe Darwin's theory of descent with modification and the mechanism, natural selection, that he proposed to explain how this comes about.*

Four student answers to the question are given. Based on what you know about Darwinian evolution and natural selection, evaluate and grade how well each answer represents Darwin's ideas. For any answer that does not receive full credit (4 points) be sure to indicate why points were lost.

Student 1. Darwin saw that populations increased faster than the ability of the land to support them could increase, so that individuals must struggle for limited resources. He proposed that individuals with some inborn advantage over others would have a better chance of surviving and reproducing offspring and so be naturally selected. As time passes, these advantageous characteristics accumulate and change the species into a new species.

Grade:

Student 2. Darwin's theory of evolution explains how new species arise from already existing ones. In his mechanism of natural selection, organisms with favorable traits tend to survive and reproduce more successfully, while those that lack the traits do not. Beneficial traits are passed on to future generations in this manner, and a new species will be created in the end!

Grade:

Student 3. Descent with modification using natural selection was Darwin's attempt at explaining evolution. An organism is modified by its surroundings, activities, and lifestyle. These modifications, by natural selection, make the organism better suited to its life.

Grade:

Student 4. Darwin's theory states that organisms can become modified by environmental conditions or use or disuse features and that the modifications can be passed down to succeeding generations. He proposes that nature selects for a characteristic trait that is beneficial to the survival of the organisms and that organisms would pass on this trait.

Grade:

Activity 23.1 A Quick Review of Hardy-Weinberg Population Genetics

Part A. Review Chapter 23 of *Biology*, 8th edition. Then complete the discussion by filling in the missing information.

If evolution can be defined as a change in gene (or more precisely, allele) frequencies, is it conversely true that a population not undergoing evolution should maintain a stable gene frequency from generation to generation? This was the question that Hardy and Weinberg answered independently.

1. **Definitions.** Complete these definitions or ideas that are central to understanding the Hardy-Weinberg theorem.

    a. Population: An interbreeding group of individuals of the same _____.
    b. Gene pool: All the alleles contained in the gametes of all the individuals in the

        _____.

    c. Genetic drift: Evolution (defined as a change in allele frequencies) that occurs in _____ populations as a result of chance events.

2. **The Hardy-Weinberg theorem.** The Hardy-Weinberg theorem states that in a population that _____ (is/is not) evolving, the allele frequencies and genotype frequencies remain constant from one generation to another.

3. **Assumptions.** The assumptions required for the theorem to be true are listed on page 472 of *Biology*, 8th edition, and are presented here in shortened form.

    a. The population is very _____.
    b. There is no net _____ of individuals into or out of the population.
    c. There is no net _____; that is, the forward and backward mutation rates for alleles are the same. For example, *A* goes to *a* as often as *a* goes to *A*.
    d. Mating is at _____ for the trait/gene(s) in question.
    e. There is no _____. Offspring from all possible matings for the trait/gene are equally likely to survive.

4. **The Hardy-Weinberg proof.** Consider a gene that has only two alleles, R (dominant) and r (recessive). The sum total of all R plus all r alleles equals all the alleles at this gene locus or 100% of all the alleles for that gene.

Let p = the percentage or probability of all the R alleles in the population
Let q = the percentage or probability of all the r alleles in the population

If all R + all r alleles = 100% of all the alleles, then

$$p + q = 1 \text{ (or } p = 1 - q \text{ or } q = 1 - p)$$

(*Note:* Frequencies are stated as percentages [for e.g., 50%] and their associated probabilities are stated as decimal fractions [for e.g., 0.5].)

Assume that 50% of the alleles for fur color in a population of mice are B (black) and 50% are b (brown). The fur color gene is autosomal.

a. What percentage of the gametes in the females (alone) carry the B allele? _____
b. What percentage of the gametes in the females (alone) carry the b allele? _____
c. What percentage of the gametes in the males carry the B allele? _____
d. What percentage of the gametes in the males carry the b allele? _____
e. Given the preceding case and all the Hardy-Weinberg assumptions, calculate the probabilities of the three possible genotypes (RR, Rr, and rr) occurring in all possible combinations of eggs and sperm for the population.

		Female gametes and probabilities	
		$\textcircled{R}(p)$	$\textcircled{r}(q)$
Male gametes and probabilities	$\textcircled{R}(p)$	RR (p^2)	_____ ( )
	$\textcircled{r}(q)$	_____ ( )	_____ ( )

Because the offspring types represent all possible genotypes for this gene, it follows that

$$p^2 + 2pq + q^2 = 1 \text{ or } 100\% \text{ of all genotypes for this gene}$$

Part B. Use your understanding of the Hardy-Weinberg theorem and proof to answer the questions.

1. According to the Hardy-Weinberg theorem, $p + q = 1$ and $p^2 + 2pq + q^2 = 1$. What does each of these formulas mean, and how are the formulas derived?

2. Assume a population is in Hardy-Weinberg equilibrium for a given genetic autosomal trait. What proportion of individuals in the population are heterozygous for the gene if the frequency of the recessive allele is 1%?

3. About one child in 2,500 is born with phenylketonuria (an inability to metabolize the amino acid phenylalanine). This is known to be a recessive autosomal trait.

    a. If the population is in Hardy-Weinberg equilibrium for this trait, what is the frequency of the phenylketonuria allele?

    b. What proportion of the population are carriers of the phenylketonuria allele (that is, what proportion are heterozygous)?

4. In purebred Holstein cattle, about 1 calf in 100 is spotted red rather than black. The trait is autosomal and red is a recessive to black.

    a. What is the frequency of the red alleles in the population?

    b. What is the frequency of black homozygous cattle in the population?

    c. What is the frequency of black heterozygous cattle in the population?

5. Assume that the probability of a sex-linked gene for color blindness is $0.09 = q$ and the probability of the normal allele is $0.91 = p$. This means that the probability of X chromosomes carrying the color blindness allele is 0.09 and the probability of X chromosomes carrying the normal allele is 0.91.

   a. What is the probability of having a color-blind male in the population?

   b. What is the probability of a color-blind female?

6. The ear tuft allele in chickens is autosomal and produces feathered skin projections near the ear on each side of the head. This gene is dominant and is lethal in the homozygous state. In other words, homozygous dominant embryos do not hatch from the egg. Assume that in a population of 6,000 chickens, 2,000 have no tufts and 4,000 have ear tufts. What are the frequencies of the normal versus ear tuft alleles in this population?

7. How can one determine whether or not a population is in Hardy-Weinberg equilibrium? What factors need to be considered?

8. Is it possible for a population's genotype frequencies to change from one generation to the next but for its gene (allele) frequencies to remain constant? Explain by providing an example.

23.1 Test Your Understanding

In each of the following scenarios, state which assumption of the Hardy-Weinberg Law is being violated and give the basis for your choice.

1. In a particular region of the coast, limpets (a type of mollusc) live on near-shore habitats that are uniformly made up of brown sandstone rock. The principle predators of these limpets are shorebirds. The limpets occur in two morphs, one with a light-colored shell and one with a dark-colored shell. The shorebirds hunt by sight and are able to see the light limpets on the dark sandstone more easily than the dark limpets.

2. In *Chen caerulescens*, a species of goose, the white body form (the snow goose) and the blue body form (the blue goose) occasionally coexist. In these areas of contact, white-by-white and blue-by-blue matings are much more common than white-by-blue matings.

3. Prior to the Mongolian invasions that occurred between the 6th and 16th centuries, the frequency of blood type B across Europe was close to zero. The frequency of blood type B among the Mongols was relatively high. Today, it is possible to see fairly high frequencies of blood type B in the Eastern European countries and a gradual decrease in the frequency of blood type B as one moves from the Eastern European countries to the Western European countries such as France and England.

Name_____   Course/Section_____

Date_____   Professor/TA_____

Activity 23.2 What effects can selection have on populations?

1. What effects can natural selection have on populations? For example, what types of selection can occur in a population, and how does each affect a population?

2. Examine the scenarios on the following pages. For each scenario:

   a. Decide whether or not natural selection is operating. In doing this, indicate whether there is variability in the population(s).

      If *no*, what does this imply about evolution?
      If *yes*, what is the nature of the variation? For example, what characteristics must the variation have for selection to operate on it?

   b. Is there any indication that members of the population(s) differ in fitness?

      If *no,* what does this imply about the operation of natural selection?
      If *yes,* describe the difference in fitness.

   c. Given your answers to parts a and b, what trend should characterize the future behavior or composition of the population(s)?

Be sure to indicate any assumptions you make in answering the questions.

Scenario I. A particular species of mouse feeds on the seeds of a single species of cherry tree. When the mice eat a seed, they digest it completely. The mice choose seeds of intermediate and large sizes, leaving the very small seeds of the cherry tree uneaten.

a.

b.

c.

Scenario IIa. Small island A contains three separate populations of a single species of cherry tree. The seed size varies between trees. That is, some trees produce seeds that are all in the small size ranges, others produce seeds all in the middle size ranges, and others produce seeds in the large size ranges. A small population of mice is introduced to the island. The mice eat cherries and are the only predators on the cherry trees. When the mice eat a cherry, they completely digest it and the pit or seed inside it. The mice choose medium and large seeds and leave the smallest seeds uneaten.

a.

b.

c.

Scenario IIb. Would your answers for Scenario IIa change given the following information? Explain. As you continue to study the populations of trees, you note that the viability of the seeds varies with size such that the viability of the small seeds is less than that of the middle-sized seeds, which is less than that of the largest seeds.

a.

b.

c.

Scenario IIIa. Small island B contains three separate populations of a single species of cherry tree. Unlike the species on island A, in this species seed size varies within trees. That is, each tree produces seeds that range in size from large to small. A small population of mice is introduced to the island. These are the only predators on the cherry trees. When the mice eat a cherry, they digest it and the pit or seed completely. The mice choose medium and large seeds and leave the smallest seeds uneaten.

a.

b.

c.

Scenario IIIb. Would your answers for Scenario IIIa change given the following information? Explain. As you continue to study these populations, you note that the viability of the seeds does not vary with size. Over time, however, you find that the trees that grow from the smallest seeds produce fewer large seeds.

a.

b.

c.

Scenario IV. After a severe spring ice storm, about half of the finches (small birds) in a population are found dead. Examination of the dead birds indicates that they vary in age from young to old. About 60% of the dead are new fledglings (just left the nest); about 20% are over 3 years of age (old for this species).

a.

b.

c.

23.2 Test Your Understanding

The figure below shows the distribution of a phenotypic characteristic in a population of organisms. Also shown are three portions (A, B, and C) of that distribution.

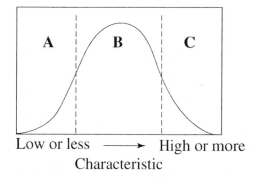

Low or less ———→ High or more
Characteristic

For the situations described in questions 1 to 3 below, which portion(s) of the population will be *selected against* and *least likely* to have their genes represented in the next generation? Explain your answers.

Use the following set of answers:
  A. Portion A only
  B. Portion B only
  C. Portion C only
  D. Portions A and C simultaneously
  E. Portions B and C simultaneously

1. Male sticklebacks with bright red coloring are favored by female sticklebacks as mates. However, the bright red color makes the males more likely to be seen by predators.

2. On Island Z in the Galápagos, the plant population contains only two species. One of the two plant species produces very large seeds and the other produces small seeds. A species of seed-eating finch has lived on the island for many years. This established species has large beaks and prefers large seeds. A small population of a different species of seed-eating finch has migrated to the island. The beak size in the new species be affected varies over a continuum from relatively small to large. *How will evolution of the new finch species be affected.*

3. In a species of plant-eating land snail, the shell color is variable, ranging from very light to intermediate to very dark. The snails are preyed upon by birds that use sight to find their prey. A small population of these snails is moved to an island where the food plants in their preferred habitat are either very light in color or very dark.

Activity 24.1 What factors affect speciation?

1. The Galápagos Archipelago consists of a dozen islands, all within 64 km of their nearest neighbor. From 1 to 11 of the 13 species of Darwin's finches live on each island. Many evolutionary biologists believe that if there had been only one island, there would be only one species of finch. This view is supported by the fact that Cocos Island is isolated (by several hundred kilometers of open ocean) from the other islands in the archipelago and only one species of finch is found there.

   a. How does the existence of an archipelago promote speciation? Explain and provide an example.

   b. Is the mode of speciation that occurred on these islands more likely to have been allopatric or sympatric? Explain.

   c. Is the type of speciation seen on the Galápagos Archipelago more likely to be the result of anagenesis or cladogenesis? Explain.

2. Hybrids formed by mating two different species are often incapable of reproducing successfully with each other or with the members of their parent populations. Explain why this is the case. (*Hint:* Consider what you know about chromosome numbers and meiosis.)

3. Because most hybrids can't reproduce, their genes (and the genes of their parents) are removed from the population. Only the genes of individuals who breed with members of their own species remain in the population. This implies that there is a strong selective advantage for genes that enable individual organisms to recognize members of their own species. Today a wide range of reproductive isolating mechanisms has been identified.

   Each of the following scenarios describes a reproductive isolating mechanism. Indicate whether each is a prezygotic or postzygotic isolating mechanism. Explain your answers.

   a. Crickets use species-specific chirp patterns to identify a mate of their own species.

   b. Two species of butterfly mate where their ranges overlap and produce fertile offspring, but the hybrids are less viable than the parental forms.

   c. Two species of a plant cannot interbreed because their flowers differ in size and shape and require pollination by different species of bee.

   d. Two species of firefly occupy the same prairie and have similar flash patterns, but one is active for a half-hour around sunset while the other doesn't become active until an hour after sunset.

4. Many of our most successful grain crops arose as hybrids; most are also allopolyploids. These crops can successfully reproduce. Explain.

Name________________________ Course/Section________________________

Date________________________ Professor/TA________________________

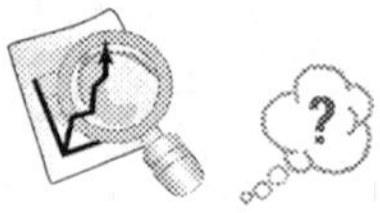

Activity 24.2 How does hybridization affect speciation?

As noted in the text, two species of toads, *Bombina bombinna* and *Bombina variegata,* share a hybrid zone that is 4,000 km in length and only 10 km wide. The frequency of alleles specific to *B. bombina* decreases from close to 100% on one edge of the hybrid zone to 0% on the opposite edge. Similarly, the frequency of *B. variegata*–specific alleles decreases across the hybrid zone (beginning at the opposite edge) from close to 100% to 0%.

1. On the graph below, map the general percentage of each type of species-specific genes across the hybrid zone. Use an X to indicate the frequency of *B. bombina*–specific genes and an O to indicate the frequency of *B. variegata*–specific genes.

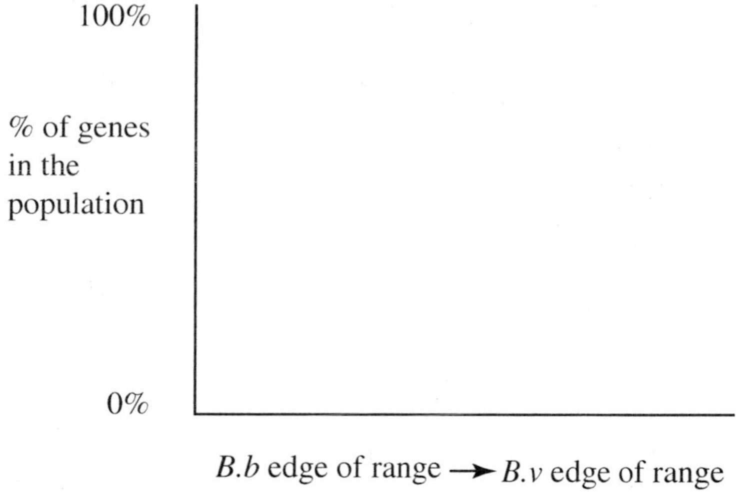

100%
% of genes in the population
0%
B.b edge of range → B.v edge of range

2. How would you predict this gene distribution would change over time if:

a. reinforcement occurs?

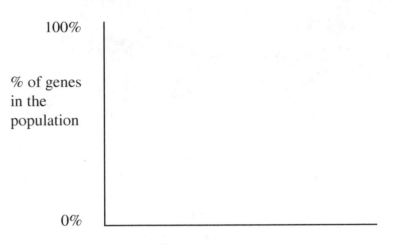

B.b edge of range ⟶ B.v edge of range

b. fusion occurs?

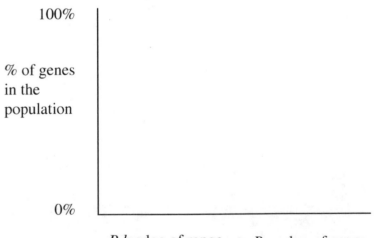

B.b edge of range ⟶ B.v edge of range

c. Hypothetically, which (a or b) is more likely to occur if all environmental conditions across the two species ranges are similar? Explain.

d. Hypothetically, which (a or b) is more likely to occur if the environmental conditions vary gradually across the species ranges such that one end of the range is, for example, much colder than the other? Explain.

e. As noted in the text, what changes in allele frequencies have been recorded over this hybrid range in the last 20 years? What does this indicate?

24.2 Test Your Understanding

Read the following statement. Then, based on your knowledge of cell biology, genetics, and evolution, decide to agree or disagree with the statement. Whichever you decide, write a short paragraph that provides solid evidence defending your choice.

"When you think about sexual reproduction, it makes no sense. After all, evolution selects for organisms that are best fit. In a population of sexually reproducing organisms, a mutant that reproduced asexually would increase its representation more quickly than the wild type; that is, it would have a higher fitness. So we should expect asexual reproduction to be much more widespread among eukaryotic species than it is."

Name_____   Course/Section_____

Date_____   Professor/TA_____

Activity 25.1 What do we know about the origin of life on Earth?

Construct a concept map of conditions on the early Earth and the origin of life-forms. Be sure to include definitions or descriptions of all the terms in the list below. Keep in mind that there are many ways to construct a concept map.

- Begin by writing each term on a separate sticky note or piece of paper.
- Then organize the terms into a map that indicates how the terms are associated or related.
- Draw lines between terms and add action phrases to the lines that indicate how the terms are related.
- If you are doing this activity in small groups in class, explain your map to another group of students when you are done.

Here is an example:

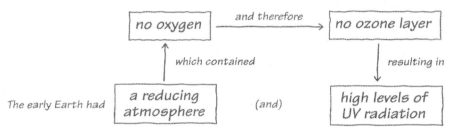

Terms

no oxygen	protobiont	prokaryotes
reducing atmosphere	micelle	RNA world
high-oxygen atmosphere	phospholipid bilayer	Eukarya
sunlight	ammonia	energy source
electrical discharge (for example, lightning)	phospholipids	carbon source
amino acids	water	mode of nutrition
ozone layer	Stanley Miller	anaerobic bacteria
sugars	methane	cyanobacteria (blue-green algae)
nucleic acids	molecular clocks	
DNA	heat	Gram stain
inorganic compounds	heterotrophs	antibiotics
organic compounds	autotrophs	penicillin
carbon dioxide	high levels of UV radiation	
	low levels of UV radiation	

Use the understanding you gained from creating the concept map to answer the questions.

1. Modern theory suggests that the early (pre-life) atmosphere on Earth was a reducing one. Why (for what reasons) is it believed that oxygen was not present when life formed on Earth?

2. What proposed energy sources existed on this early (pre-life) Earth?

3. In the 1950s, Stanley Miller performed a set of experiments to determine whether life could have evolved given the conditions stated in the answers to questions 1 and 2.

   a. How was the experiment designed?

   b. What were the necessary controls?

c. What was produced in the experiment?

d. What did the results imply about the possible origin of life on Earth?

e. There is general agreement that life must have evolved in the oceans originally and only much later invaded land. What factors of the physical environment on the early and evolving Earth support these ideas? Changes in which of these factors were essential for life to survive on land?

f. Most of us can't imagine a world without oxygen. However, as you learned earlier, chemically oxygen is a powerful oxidizing compound. What effect(s) would the increase in oxygen levels of the atmosphere have on the organisms that existed at that time?

Activity 25.2 How do we determine the age of fossils and rocks?

To determine the age of fossils and rocks, scientists determine the amounts of radioactive compounds and their stable daughter products present in the sample. Radioactive elements are known to decay into stable daughter compounds at specific rates. A number of radioactive compounds, their stable daughter compounds, and their half-lives are shown in the table below.

Radioactive compound	Stable daughter compound	Half-life
Carbon 14	Nitrogen 14	5370 years
Potassium 40	Argon 40	1.25 billion years
Rubidium 87	Strontium 87	48.8 billion years
Thorium 232	Lead 208	14 billion years
Uranium 235	Lead 207	704 million years
Uranium 238	Lead 206	4.47 billion years

For dating rocks, the potassium-argon method is often used because:

- Argon is a gas. When rock is molten, any existing argon gas will escape. As a result, newly formed rocks contain no argon.
- As the ^{40}potassium naturally occurring in the rock decays to ^{40}argon, the ^{40}argon is trapped in pockets in the rock. The ratio of ^{40}potassium to ^{40}argon can be measured to determine the rock's age.

For dating organic material, carbon-nitrogen dating is often used because:

- The ratio of radioactive to nonradioactive carbon dioxide in the atmosphere is fairly constant over time. As a result, the levels of these elements in organic tissue remain relatively constant as long as the organism is alive.
- Once the organism is dead, no new inputs of carbon occur and the existing radioactive-carbon–to–nonradioactive-carbon ratio will decrease over time as the radioactive carbon decays.

1. In one half-life, half of the original radioactive compound will decay into its stable daughter compound.

   a. If a newly formed rock contains 100 units of 40 potassium (^{40}K), how many units of potassium 40 (^{40}K) would it contain after 1.25 billion years?

   b. How many units of argon 40 (^{40}Ar) would the rock contain when newly formed vs. after 1.25 billion years?

c. After 1.25 billion years, what would be the ratio of ^{40}K to ^{40}Ar in the rock?

d. After 2.5 billion years, what would be the ratio of ^{40}K to ^{40}Ar?

2. You are fossil hunting and find a trilobite fossil in an old riverbed. You have it radiometrically dated and are told it is 275 million years old.

   a. In this amount of time, how many half-lives of ^{40}K have elapsed?

   b. Given your answer in a, what would be the ratio of ^{40}K to ^{40}Ar found in the fossil remains?

3. You want to date some fabric that you have discovered at an archeological dig.

   a. What method of dating would be best for this? Describe the general procedure for dating the cloth.

   b. Assume the cloth is 2,000 years old. How would the level of the radioisotope you used to date it have changed in this period of time?

Name_____   Course/Section_____

Date_____   Professor/TA_____

Activity 26.1 How are phylogenies constructed?

Construct a modified concept map to relate the concepts of phylogeny and systematics listed below to the phylogenetic tree on the next page.

- Begin by writing each term on a separate sticky note or piece of paper.
- Then indicate how the terms are associated or related to each other and to the phylogenetic tree on the next page.
- Be sure to include definitions or descriptions of all the terms as you use them to explain these relationships.

Terms

clade	monophyletic	shared derived character
cladistics	polyphyletic	outgroup
phylogenetic tree	paraphyletic	ingroup
homology	convergent evolution	taxonomy
analogy	shared primitive character	phylogeny

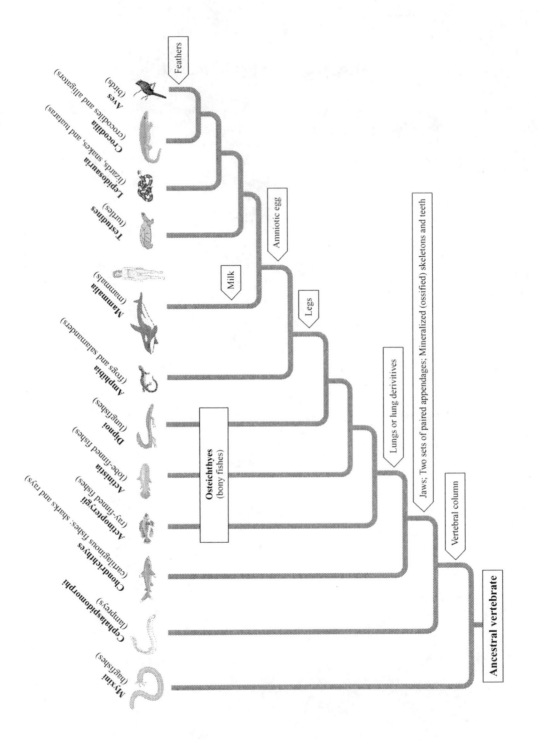

Feathers

Aves
(birds)

Crocodilia
(crocodiles and alligators)

Lepidosauria
(lizards, snakes, and tuataras)

Testudines
(turtles)

Ammiotic egg

Mammalia
(mammals)

Milk

Amphibia
(frogs and salamanders)

Legs

Dipnoi
(lungfishes)

Actinistia
(lobe-finned fishes)

Lungs or lung derivitives

Osteichthyes
(bony fishes)

Actinopterygii
(ray-finned fishes)

Jaws; Two sets of paired appendages; Mineralized (ossified) skeletons and teeth

Chondrichthyes
(cartilaginous fishes: sharks and rays)

Cephalaspidomorphi
(lampreys)

Vertebral column

Myxini
(hagfishes)

Ancestral vertebrate

Use the understanding you gained from creating the concept map to answer the questions.

1. Compare the taxonomy of a group with its phylogeny (in general terms).

	Taxonomy	Phylogeny
a. Definition or purpose		
b. Types of characters used to develop		
c. What similarities could there be between the taxonomy of a given group and its phylogeny?		
d. What are the key differences between the taxonomy of a given group and its phylogeny?		

2. On the phylogenetic tree shown earlier, are the groups that contain humans, whales, crocodiles, and birds monophyletic, polyphyletic, or paraphyletic? Explain.

3. Considering only the individual representative organisms in the phylogenetic tree (e.g., bird, whale, frog), which can be used as good examples of analogy, or convergent evolution? As good examples of homology? Explain your reasoning.

4. In recent years, DNA sequence analysis has been used in developing phylogenetic relationships among organisms.

   a. What type of DNA has been used most commonly in this analysis? Why was this type chosen over others?

   b. The phylogenies developed using DNA sequence analysis may differ from those constructed using morphology and physiology. How do scientists know which method is more correct?

5. Based on DNA sequence analysis, three major domains of life have been proposed
   What are the three major domains of life? What sets of characteristics place
   organisms into one domain versus another?

Major domains of life	Key characteristics

Name_____   Course/Section_____

Date_____   Professor/TA_____

Activity 26.2 What is parsimony analysis?

To determine relatedness among species, parsimony analysis is often used. In Figure 26.15 of *Biology*, 8th edition, parsimony analysis is applied to matched DNA sequences from three species of birds. The same type of analysis can be applied to morphological and developmental characteristics of species. Review Figure 26.15 and the parsimony method used. Then conduct your own parsimony analysis using the morphological characteristics of these birds in the table below.

1. Begin by recording in the table below the morphological characteristics of the three bird species in Figure 26.15.

2. On the three possible phylogenies for these species (below the table), indicate the number of changes that must have occurred for each of the proposed phylogenies to be correct.

Traits	A	B	C	D	E
Species	Eye ring (light/dark)	Red bar on breast (yes/no)	Red wing bar (yes/no)	Light bar on tail tip (yes/no)	Beak shape (curved/point)
I					
II					
III					
Ancestral					

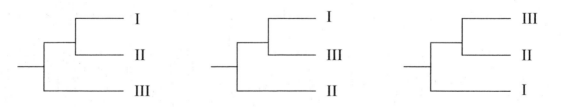

3. Based on your analysis, which of the phylogenies is most parsimonious? How does this result compare to the result given in Figure 26.15?

4. Which of the proposed phylogenies (the one you developed or the one in Figure 26.15) is more correct? Explain your answer.

Name_____   Course/Section_____

Date_____   Professor/TA_____

Activity 26.3 Put yourself in the professor's shoes: What questions would you ask?

One of the best ways to study for an exam is to put yourself in the professor's shoes. For example, ask yourself: What questions would I ask about the material if I were Professor _____? Asking and answering such questions are good practice for taking the actual exam. They also help you to better understand and organize the major ideas and concepts you have studied.

Write three exam questions designed to test how well a student understands the major concepts in Chapters 22–26. Indicate the correct answer to each question and also tell the reason why each alternative answer is incorrect. Your questions should be of the following types:

I. Problem solving or application of a concept or principle to a problem

For example: The little-known hypothetical organism *Skyscra parius* is a long-necked animal that feeds on the leaves of Australian trees that grow to heights of 30 feet. Being a hooved animal, *S. parius* cannot climb trees, so it feeds much like modern-day giraffes do. Fossil evidence indicates that the ancestors of *S. parius* had fairly short necks. Read the arguments presented below. For each, indicate whether or not the factors described could have affected neck length or tree height over the course of evolution of *S. parius*. [A = factors described could have affected neck length; B = factors described probably did not affect neck length]

_____ 1. *S. parius* ancestors likely demonstrated significant variation in neck length, with some having shorter necks and others having longer necks.
_____ 2. When first born, juveniles of *S. parius* were much shorter than adults, so they were not able to compete successfully with adults that had longer necks.
_____ 3. Female *S. parius* preferentially mate with longer-necked males.

II. Translation: the ability to recognize concepts restated in a different form or to restate concepts in a different form

For example: In what ways are the structure and function of the angiosperm seed and the amniotic egg (in this example, the chicken's egg) similar? In what ways are they different?

T/F   1.   Both the angiosperm seed and the chicken's egg contain stored food for the early development of the embryo.

T/F   2.   When released from the plant or laid by the chicken, both the angiosperm seed and the chicken's egg contain a partially developed diploid embryo/offspring generation.

T/F   3.   The seed coat of the seed and the shell of the egg help prevent desiccation (water loss).

T/F   4.   Seeds may remain dormant and viable for hundreds of years; the same is true of chicken eggs.

Write your exam questions in the spaces provided.

Exam Question 1:

Answer:

Exam Question 2:

Answer:

Exam Question 3:

Answer:

Activity 26.3

Name_____   Course/Section_____

Date_____   Professor/TA_____

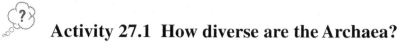 **Activity 27.1 How diverse are the Archaea?**

1. The Archaea are divided into two major groups: the Euryarchaeota and the
    Crenarchaeota.

    a. What characteristics are used to place organisms into each of these groups?

Archaean group	Characteristics
Euryarchaeota	
Crenarchaeota	

    b. Two additional groups of the Archaea have been proposed.

What new groups are these?	What characteristics do members of these groups have?

2. There is great diversity in the ways different species of microbes

  • obtain energy for metabolic functions, and
  • obtain carbon for building the macromolecules of life.

Fill in the chart to indicate how these characteristics are used to describe the nutritional type or nutritional classification of organisms. (Refer also to Table 27.1 on page 564 in *Biology,* 8th edition.)

	Plants	Animals	Bacteria and/or Archaea					
Energy source	*Light*	*Organic compounds*	*Light*	*Organic compounds*	*Light*	*Organic compounds*
Carbon source	*Inorganic CO$_2$*	*Organic compounds*				
Mode of nutrition	*Autotroph*	*Heterotroph*				

Name_____   Course/Section_____

1. Bacteria first appear in the fossil record about 3.5 billion years ago. Humans first appear in the fossil record only a few million years ago. Given this, which group would you say is more highly evolved?

    a. What kinds of arguments or evidence would you use to support the idea that bacteria are more highly evolved?

    b. How would you support the idea that humans are more highly evolved?

2. Given what you know about the origin of life on Earth, you want to look for life on other planets.

    a. What characteristics of the planet's environment would you look for? Explain your reasoning.

b. What kind(s) of life would you look for? Explain your reasoning.

c. What tests or probes would you use to find the kind of life you proposed in question b? Explain your reasoning.

Activity 27.2 How has small size affected prokaryotic diversity?

It is often said that bacteria tend to be more diverse biochemically and eukaryotes tend to be more diverse morphologically.

Answer questions 1–7. Then write a summary argument (question 8) to support the statement above. In other words, write an argument describing how small size has limited morphological diversity but promoted biochemical (metabolic) diversity among the prokaryotes.

1. All living organisms must maintain a relatively constant internal environment. Maintaining this environment means that a certain concentration of each substance must be maintained per unit volume of the cell. The ability of the cell to maintain a specific concentration of a substance is affected by

    • the ability of the substance to diffuse through the membrane,
    • the overall size and shape of the cell, and
    • the maximum amount of the substance a given area of membrane can transport per unit time.

    a. How do surface-area-to-volume (*SA/V*) ratios change as the size and shape of cells and organisms change? To answer this, calculate the *SA* and *V* of a cube 1 mm on a side. Then do the same for cubes that are 2 mm and 4 mm on a side and compare their *SA/V* ratios.

Cubes	1-mm square	2-mm square	4-mm square
Linear dimension	1 mm	2 mm	4 mm
Surface area (*SA*)	$6(1 \text{ mm}^2) = 6 \text{ mm}^2$		
Volume (*V*)	$(1 \text{ mm})^3$		
SA/V ratio	$6 \text{ mm}^2/1 \text{ mm}^3$		

    b. In general, how does surface area change as linear dimensions increase twofold?

c. In general, how does volume change as linear dimensions increase twofold?

d. In general, how do *SA/V* ratios change as linear dimensions increase twofold?

2. Assume a bacterium is 10 μm in linear dimension. Fill in the chart.

   a. If modeled as a cube, what would its *SA, V,* and *SA/V* ratio be?
   b. If modeled as a sphere, what would its *SA, V,* and *SA/V* ratio be?
   c. What are the *SA* and *V* values and the *SA/V* ratios for a cube-shaped eukaryotic cell that is 100 μm in linear dimension?

	a. 10-μm bacterium as a cube	b. 10-μm bacterium as a sphere	c. 100-μm eukaryote, cube-shaped
SA			
V			
SA/V ratio			

3. Assume that every cell requires a minimum of 1 unit of oxygen per μm^3 per second to stay alive. Fill in the chart.

   a. How much oxygen must cross each μm^2 of surface area per second in the 10-μm bacterium versus the 100-μm eukaryote to keep each alive?
   b. What effects might this difference have on metabolic rates in these organisms?

	10-μm bacterium	100-μm eukaryote
a. Oxygen/μm^2 of *SA*/second		
b. Possible effect(s) on metabolic rate		

4. Given what you know about cell membranes, is there likely to be a maximum upper limit on the number of molecules of a substance that can cross a given area of membrane per unit time? If so, what factors would be involved in determining the maximum upper limit?

As you answer questions 5, 6, and 7, fill in the chart on the next page.

5. a. On average, how large is a prokaryotic genome?

   b. On average, how many times larger is a eukaryotic genome?

   c. Are these genomes haploid or diploid?

6. Under ideal conditions, many bacterial cells can reproduce or duplicate themselves within an hour. Some species (for example, *E. coli*) can reproduce every 20 minutes under ideal conditions. If you inoculate 1 liter of culture medium with one bacterium per milliliter (of medium) at time $t = 0$, how many bacteria will be present in 1 milliliter of the culture medium after 10 hours? (To do this calculation, assume that in these culture conditions the bacteria duplicate—or double in number—every hour.)

7. Assume that the mutation rate of bacterial cells in culture is 10^{-6} to 10^{-8}. This means that in every 1 million to 100 million cells produced from a single original cell, you would expect to find at least one mutation. How many mutations would you be likely to find in 1 liter of the 10-hour culture of the cells you grew in question 6?

	Prokaryote	Eukaryote
Genome size	5a.	b.
5. Haploid versus diploid		
6. Maximum reproduction rate		One hour to many hours
7. Mutation rate per hour		

8. Using all of the information in this activity, write an argument entitled: "How small size in prokaryotes played a role in limiting their morphological diversity and promoting their biochemical (metabolic) diversity."

Name_____   Course/Section_____

Date_____   Professor/TA_____

Activity 28.1 How has endosymbiosis contributed to the diversity of organisms on Earth today?

1. The protists have been called a "catch-all group." What does this mean? Explain.

2. There are more than 100,000 recognized species of protists; in *Biology*, 8th edition, they are subdivided into five supergroups and 11 major clades. Even though the protists are extremely diverse, we can recognize some common themes in their evolution, including the evolution of complex cell structure, novel genetic recombination strategies, and complex life cycles. In addition, this group as a whole has great ecological importance. Provide at least two examples of protists from Chapter 28 in *Biology*, 8th edition, that illustrate each theme.

Theme	Examples
a. Complex cell structure	
b. Novel genetic recombination strategies	
c. Complex life cycles	
d. Ecological importance	

3. In the late 1960s (and since), Lynn Margulis provided considerable evidence for the endosymbiotic theory of the origin of various organelles in eukaryotic cells.

a. What is the endosymbiotic theory?	
b. Which two eukaryotic organelles were proposed to have arisen as endosymbionts?	c. What evidence did Margulis present to support each organelle as an endosymbiont?

4. The mitochondria of the chromalveolates and the chlorarachinophytes (in the clade Rhizaria) are both thought to have arisen as a result of secondary endosymbiosis.

    a. What is secondary endosymbiosis?

    b. What evidence supports the idea of secondary endosymbiosis?

5. Structurally, *Giardia lamblia* lacks complete mitochondria and was once thought to be an example of what the earliest living eukaryotes may have looked like.

    a. If *Giardia* is similar in structure to the earliest living eukaryotes, what does this imply about the order of evolution of the various eukaryotic organelles (that is, nucleus, cytoskeleton, mitochondria, chloroplasts)?

    b. More recently, we have discovered that the Diplomonads (e.g., *Giardia*) and the Parabasalids (the other group of Excavates) have modified mitochondria.

        i. What types of mitochondria do they have?
        ii. What does this evidence indicate about the evolution of the various eukaryotic organelles?

28.1 Test Your Understanding

The metabolic pathways of organisms living today evolved over a long period of time—undoubtedly in a stepwise fashion because of their complexity. Considering everything you have learned to date about the evolution of life on Earth, put the following in the order in which they might have evolved, and provide an explanation for your arrangement.

____ prokaryotes capable of performing the Krebs cycle
____ eukaryotes capable of performing the Krebs cycle
____ prokaryotes capable of performing electron transport
____ eukaryotes capable of performing electron transport
____ prokaryotes capable of performing glycolysis
____ eukaryotes capable of performing glycolysis
____ prokaryotes capable of performing photosynthesis
____ eukaryotes capable of performing photosynthesis

Name_____   Course/Section_____

Date_____   Professor/TA_____

 Activity 29.1/30.1 What major events occurred in the evolution of the plant kingdom?

Construct a concept map that describes the early evolution of plant life on Earth. Be sure to include relationships among all the organisms and factors in the list on the next page. Keep in mind that there are many ways to construct a concept map.

- Begin by writing each term on a separate sticky note or piece of paper.
- Then organize the terms into a map that indicates how the terms are associated or related.
- Draw lines between terms and add action phrases to the lines that indicate how the terms are related.

Here is an example:

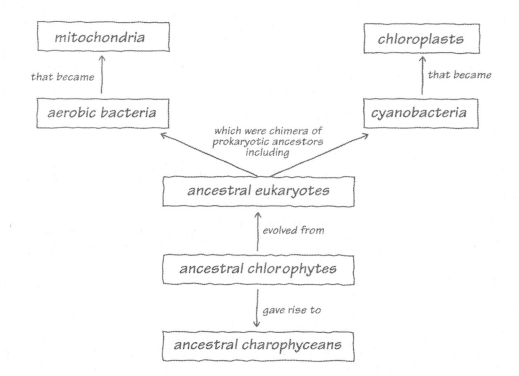

If you are doing this activity in small groups in class, explain your map to another group of students when you finish it.

Terms

bryophytes	alternation of generations	spore
pteridophytes	megaspore	gametophyte
lycophytes	microspore	sporophyte
endosymbiont	nonvascular plants	egg
anaerobic bacteria	seedless	gametangia
cyanobacteria (blue-green algae)	seeds	root
	angiosperm	stem
chloroplast	gymnosperm	flagellated sperm
mitochondria	flowers	archegonium
vascular tissue	xylem	antheridium
waxy cuticle	phloem	pollen grain
charophyceans	microphyll	ovule
chlorophytes	megaphyll	

Use the understanding you gained from constructing the concept map to answer the questions.

1. Describe the major problems ancestral land plants had to overcome before they could make the transition from water to land.

2. Describe the major solutions to the problems in question 1 that can be found in today's land plants. In other words, what mutations occurred that allowed organisms to make the transition to land?

Name_____   Course/Section_____

Date_____   Professor/TA_____

Activity 29.2/30.2 What can a study of extant species tell us about the evolution of form and function in the plant kingdom?

Fill in the chart on the next pages to compare the major features of key groups of land plants with one another and with the charophyceans.

Plant Group

Feature	Bryophytes	Lycophytes	Pterophytes	Gymnosperms	Angiosperms	Charophyceans (green algae)
Peroxisomes	Yes	Yes	Yes	Yes	Yes	Yes
Chlorophylls *a* and *b*	Yes	Yes	Yes	Yes	Yes	Yes
Jacketed gametangia	Yes	Yes	Yes	Yes	Yes	No
Cuticle						
Stomata						
Vascular tissue						
Stems containing vascular tissue						
Roots or rhizomes						
True leaves (contain vascular tissue)						
Antheridia						
Archegonia						
Flagellate sperm						

Plant Group

Feature	Bryophytes	Lycophytes	Pterophytes	Gymnosperms	Angiosperms	Charophyceans (green algae)
Pollen						
Seed						
Flower						
Fruit						
Gametophyte dominant						
Sporophyte dominant						
Sporophyte dependent on gametophyte for energy						
Sporophyte and gametophyte both independent						
Gametophyte dependent on sporophyte						

Use the information in the chart you have completed to answer the questions.

1. Some of the major plant groups are listed in the following table from most primitive to most advanced. For each group, indicate what major characteristics make it more advanced than the preceding group. For example, how are ferns more advanced than mosses?

	Plant group					
	Charophyceans	Bryophytes	Lycophytes	Pterophytes	Gymnosperms	Angiosperms
Advance(s) over preceding group						

2. How do the bryophytes differ from the seedless vascular plants? How are they similar?

3. The life cycle of all land plants includes an alternation of generations between a multicellular gametophyte phase and a multicellular sporophyte phase.

   Diagram the life cycle of a seed plant.

   a. What cellular division process always precedes formation of the gametophyte generation?

   b. What cellular process always precedes formation of the sporophyte generation?

   c. If the sexual generation gives rise to the gametes, what part of an angiosperm is sexual?

   d. If the sexual generation gives rise to the gametes, what part of a bryophyte moss is sexual?

4. Until the evolution of the seed plants, land plants were dependent on the availability of water for reproduction. Explain why this was true. Explain how seed plants overcame the need for water in reproduction.

5. Pollen, seeds, flowers, and fruits are considered among the most advanced characteristics in the plant kingdom. What evolutionary advantage(s) does each of these offer (relative to what existed before)?

Activity 29.2/30.2

Name_____   Course/Section_____

Date_____   Professor/TA_____

Activity 29.3/30.3 How are the events in plant evolution related?

1. Working in groups of three or four, assign each student in the group one of the following events in plant evolution to research. Then give each student 5 minutes to report the results of his or her review to the other members of the group.

Events in Plant Evolution

I. Evolution of vascular tissue

Mutations in some land plants gave rise to vascular tissue. What advantage(s) did these plants have compared with land plants that did not contain any vascular tissue?

II. Evolution of roots and leaves

Mutations in some land plants gave rise to roots, leaves, or both. What advantage(s) did these plants have compared with plants that did not contain roots or leaves?

III. The trend toward reduction of the gametophyte generation

Mutations in some land plants gave rise to life cycles in which the gametophyte generation was reduced. What advantage(s) did these plants have compared with plants that did not have a reduced gametophyte generation?

IV. Evolution of the seed

Mutations in some land plants gave rise to the seed. What advantages did these plants have compared with plants that did not have seeds?

2. Work together as a group to determine how these events in plant evolution (I to IV) might be related. For example, which would have to come first (in evolution), which next, and so on? Another way to look at this question is to consider which of these events paved the way (or made it possible) for the other events to occur.

Be sure to state evidence for your proposed evolutionary scheme. To do this, it is useful to ask yourself questions like these: Would it have been possible for the seed to evolve without vascular tissue first having evolved? Leaves? And so on. If yes, how could this have occurred? If no, why not?

3. Incorporate the following *true* observations into your analysis of how the events in evolution (I to IV) could be related.

   a. The fossil record (spore evidence and so on) indicates that the first plants on Earth were similar to modern-day bryophytes.

   b. The very large marine brown algae (for example, giant kelp) can grow to heights of 30 feet or more. These algae have both leaflike and stemlike structures and are held to the bottom of the sea by a holdfast. When examined microscopically, the algae are found to contain transport vessels that are similar to phloem in function. These giant kelp do not contain any xylemlike vessels, however, nor do they have roots.

   c. The first land plants with xylem and phloem had no leaves or roots.

   d. Some seedless vascular sporophyte plants do not release megaspores. Instead, the megaspore divides on the sporophyte (in the old sporangium) to produce the female gametophyte. This female gametophyte produces eggs in archegonia, which are fertilized by sperm produced in antheridia of this or other plants. Neither the female gametophyte (once formed) nor the developing embryo receives nutrition from the old sporophyte plant.

4. Write an analysis of how events a through d could be related.

29.3/30.3 Test Your Understanding

Consider a puddle of water in which mosses, ferns, and an angiosperm are growing. In a drop of water taken from the puddle, you observe flagellated sperm.

Which of the following statements concerning the sperm could be true and which are definitely false?

Consider each statement separately and explain your answers.

1. (T/F) The sperm could have been produced by an antheridium on the moss sporophyte.

2. (T/F) The sperm could later stop swimming and develop into the male gametophyte of the fern.

3. (T/F) The sperm may be swimming to the archegonium of the fern.

4. (T/F) The sperm could have been released from a pollen tube of an angiosperm.

Name_____   Course/Section_____

Date_____   Professor/TA_____

 Activity 31.1 How diverse are the fungi in form and function?

1. a. What is the basic body plan of most fungi?

    b. Which fungi do not share this basic body plan?

2. Fungi may be said to have both plantlike and animal-like characteristics. What
    plantlike characteristics do fungi have? What animal-like characteristics?

3.

a. Into what five major phyla (or divisions) is the kingdom Fungi divided?	b. On what basis are these divisions made?

4. Is it more correct to describe a mushroom as haploid or diploid? Explain.

5. If you did not know that fungi were primarily terrestrial organisms, what structures or features of the organisms would suggest that they were terrestrial?

6. a. In what ways are fungi important in the ecosystem?

   b. In what ways are fungi important to humans?

7. Lichens are symbiotic associations of a fungus, usually an ascomycete, and an alga, usually a green algal species or a cyanobacterial species. Lichens can often survive in harsh natural environments.

   a. To what environmental conditions are lichens well adapted?

   b. What makes them so well adapted to these conditions?

Name_____   Course/Section_____

31.1 Test Your Understanding

Patients with AIDS often acquire opportunistic infections. Imagine a patient with AIDS (caused by human immunodeficiency virus) and two opportunistic infections—a respiratory disease caused by a *Mycobacterium* and another disease caused by the fungus *Candida*. As the patient's physician, you need to prescribe drugs to counteract the infections.

Given the normal anatomical and functional characteristics of each disease organism, what characteristics should a drug combination have to treat each of these three infections while doing the least harm to the patient?

Name_____ Course/Section_____

Date_____ Professor/TA_____

Activity 32.1/33.1 What can we learn about the evolution of the animal kingdom by examining modern invertebrates?

Fill in the chart on the next two pages to organize the major characteristics of key invertebrate phyla.

Key invertebrate phyla

Characteristics	Porifera	Cnidaria	Platyhelminthes	Nematoda	Annelida	Mollusca	Arthropoda	Echinodermata
Examples of organisms	*Sponges*							
Number of tissue layers in embryo	*Doesn't apply*	*2; ectoderm and endoderm*						
Tissue versus organ level development	*Quasi-tissue level*	*Tissue level*						
True muscle cells?	*No*	*No; have epithelio-muscular cells*						
Symmetry? Cephalization?								

Name_____   Course/Section_____

Key invertebrate phyla

Characteristics	Porifera	Cnidaria	Platyhelminthes	Nematoda	Annelida	Mollusca	Arthropoda	Echinodermata
Coelom? Type?								
Digestive tract? Type?								
Ciculatory system? Type?								
Nervous system? Type?								
Other								

Using the information in the chart and in Chapters 32 and 33 of *Biology*, 8th edition, answer the questions.

1. What set of characteristics is shared by all of the invertebrate animal phyla in the chart?

2. What unique combination of characteristics defines each of the invertebrate phyla as separate from the other phyla?

3. If you compare the characteristics of one phylum of the invertebrates with the next, what key differences separate the groups from each other?

4. a. Looking across the rows, what major trends appear to occur in the evolution of various organs or organ systems in these animal groups?

   b. What developmental evidence is used to link Annelids, Arthropods, and Molluscs evolutionarily?

   c. What evidence is used to separate the phylain from the Echinoderms and Chordates?

d. Does this analysis provide evidence for or against the statement: "Evolution adds onto or modifies what already exists"? Explain.

5. The chart organizes the major groups of animals based on grade, or shared body plan features. What changes would you need to make in this organization to reflect the possible phylogenetic relationships uncovered using molecular evidence? To answer this:

   a. On a separate sheet of paper, redraw the chart to reflect the new phylogenetic relationships based on molecular evidence.

   b. What specific molecular characteristics/data are being used to determine evolutionary relationships among animal phyla?

6. How would your answers to questions 2, 3, and 4 differ (if at all) when the chart is redrawn and filled in to reflect changes in relationships based on molecular evidence?

7. In biological terms, a group of organisms is said to be successful if it is represented by a large number of species or if the mass of all the organisms in the group is large. (In both cases, "large" is determined relative to other groups or organisms.) Given this definition of success, which of the major groups of animals would you argue is the most successful? Be sure to provide evidence for your argument.

Name_____   Course/Section_____

Date_____   Professor/TA_____

Activity 32.2/33.2 What factors affect the evolution of organisms as they become larger?

In the evolution of life on Earth, organisms have evolved from single celled to multicelled; small to large; simpler to more complex. Keep in mind that this apparent increase in complexity occurs because evolution adds on to or modifies what already exists.

As you discovered in Activity 27.2, surface-area-to-volume ratios and the need for organisms to gain or lose substances across their surface areas have put constraints on the evolution of cell structure and function. These same constraints affect the ability of multicellular organisms to survive. Given their small size, most of the evolution in unicellular organisms has involved changes in cell chemistry and/or internal cellular structure. In contrast, the evolution of larger and larger multicellular organisms is evidenced primarily as changes in both external and internal morphology.

1. A quick review and extension of Activity 27.2: How has small size affected prokaryotic diversity?

   a. How do surface-area-to-volume (SA/V) ratios change as the size and shape of cells and organisms change? To answer this, calculate the SA and V of a cube 1 mm on a side. Then do the same for cubes that are 2 mm and 4 mm on a side and compare their SA/V ratios.

Cubes:	1-mm square	2-mm square	4-mm square	Ratios
Linear dimension	1 mm	2 mm	4 mm	1:2:4
Surface area (SA)	$6(1 \text{ mm}^2) = 6 \text{ mm}^2$			
Volume (V)	$(1 \text{ mm})^3$			
SA/V ratio	$6 \text{ mm}^2/1 \text{ mm}^3$			

   b. What do the ratios mean?

c. Is the graph of the change (in surface area compared to volume) linear or exponential?

To answer this question, complete the following table and then graph the surface areas and volumes from this table and the previous table.

Cube linear dimension:	8 mm	16 mm	32 mm	64 mm
Surface area				
Volume				

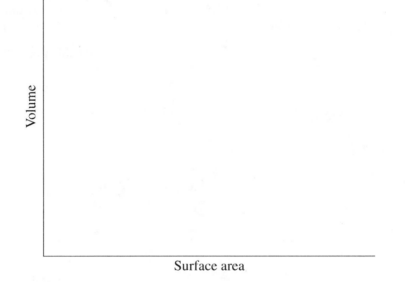

d. What effect does this relationship between surface area and volume have on the ability of larger and larger multicellular organisms to support the metabolic needs of all parts of their bodies, for example, to supply oxygen needed for cell respiration?

2. In the previous examples, as the "organism" sizes changed, their shapes remained constant (as cubes). How would changing the shape of the organism affect *SA/V* ratios? Use modeling clay or playdough to make a cube that is 2.5 cm (1 inch) on a side. The cube will obviously have a constant mass or volume.

Using your model, devise three ways to change the *SA/V* ratio of this "organism." If possible, find one way to reduce the *SA/V* ratio and two ways to increase the *SA/V* ratio. For each of your solutions to increase or decrease the *SA/V* ratio, calculate the actual change in *SA*. Enter the data you collect/calculate in the chart.

	Shape change	Effect on *SA/V* ratio
Model 1		
Model 2		
Model 3		

3. Review the chapters on fungal, animal, and plant diversity in your textbook. For each of the models that you developed to increase *SA/V* ratios, find an example of a fungus, plant, and/or animal, or a particular organ system, that uses the same method to increase *SA/V* ratios and describe it below.

Proposed model—increase *SA/V* by:	This type of increase in *SA/V* ratios can be found in the following organism(s)/organ systems:

Name_____   Course/Section_____

Date_____   Professor/TA_____

Activity 34.1 What can we learn about the evolution of the chordates by examining modern chordates?

Fill in the chart on the next two pages to organize the major characteristics of key groups of chordates.

Characteristics of phylum chordata

Characteristics	Subphylum Urochordata	Subphylum Cephalochordata	Subphylum Vertabrata						
			Agnatha	Chondrichthyes	Osteichthyes	Amphibia	Mammalia	Reptilia	Aves
Examples of organisms									
Type of skeleton									
Vertebral column?									
Jaw?									
Paired limbs? Structure? Limb position?									
Body covering?									

Characteristics of phylum chordata

Characteristics	Subphylum Urochordata	Subphylum Cephalochordata	Subphylum Vertabrata						
			Agnatha	Chondrichthyes	Osteichthyes	Amphibia	Mammalia	Reptilia	Aves
Respiratory structure?									
Circulatory system?									
Heart structure?									
Type of reproduction?									
Other									

Using the information in the table, answer the questions.

1. a. According to *Biology,* 8th edition, what three characteristics are common to all chordates?

   b. Why don't we find all three of these chordate characteristics in our own bodies?

2. The chart organizes the chordates based on grade, or shared body plan features. If you compare the characteristics of one group with the next, what key differences separate the groups from each other?

3. What unique combination of characteristics defines each group as separate from the others?

4. a. Looking across each of the rows in the chart, what major trends do you see in the evolution of the different organs and organ systems?

    b. Does this analysis provide evidence for or against the statement: "Evolution adds onto or modifies what already exists?" Explain.

5. a. What major changes in structure and function are seen in terrestrial groups as compared with aquatic groups?

    b. Can these changes be related to differences in natural selection on land versus in an aquatic environment? Explain.

Name_____   Course/Section_____

Date_____   Professor/TA_____

Activity 35.1 How does plant structure differ among monocots, herbaceous dicots, and woody dicots?

The chart on the next page shows a drawing of a generic plant. The arrows indicate various plant parts or organs. In the columns to the right of the plant, draw cross sections of the plant at the points indicated by the arrows.

To help you visualize the differences in structure among different types of plants, in Column I draw the cross sections assuming the plant is a monocot. In Column II draw the cross sections assuming the plant is a herbaceous dicot. Finally, in Column III draw the cross sections as if the plant is a woody dicot. Be sure to label your drawings.

Generic plant	Column I Cross sections if plant is a monocot	Column II Cross sections if plant is a herbaceous dicot	Column III Cross sections if plant is a woody dicot
a — b — c — d — Soil surface — e — f —			

Use the information in your drawings to answer the questions on the next pages.

Activity 35.1

Name_____   Course/Section_____

1. In Column I, connect the cross sections of the monocot by drawing a line (color 1) from the water transport tissue in one cross section to the water transport tissue in the next, and so on. Draw another line (color 2) to connect the food transport tissue from one cross section to the next. Do the same for the herbaceous and the woody dicot cross sections. How do the positions of these tissues change from one cross section to the next?

2. Use the information in the cross sections to fill in the chart below. *Note:* A distinguishing feature is one that is found in only the given type of organism. For example, a distinguishing feature of mammals is the presence of hair: Hair is a characteristic of all mammals and is not found in any other animals.

	Distinguishing feature(s) of			
	Leaf	Stem	Root	Apical meristem shoot vs. root
Monocot				
Herbaceous dicot				
Woody dicot				

3. If cross sections were not available, what other characteristics of the plants as a whole could you use to determine whether each was a monocot, a herbaceous dicot, or a woody dicot?

4. A cartoon shows a man going to sleep in a hammock suspended between two relatively short trees. The second frame of the cartoon shows the man waking 20 years later and finding his hammock 15 feet higher off the ground. Critique this drawing in terms of what you know about the growth pattern of trees.

35.1 Test Your Understanding

How is the general morphology of the various organs of a plant correlated with the function(s) of those organs? For example, why are leaves generally thin and flat? What structural advantage is provided by stems having their vascular tissue arranged in a ring near the periphery of the stem? What characteristics of roots and root growth dictate that the vascular tissue be more centrally located?

Name_____   Course/Section_____

Date_____   Professor/TA_____

Activity 36.1 How are water and food transported in plants?

Using the cross sections of the plants you developed in Activity 35.1 you identified the location of water-conducting xylem in roots, stems, and leaves. In this activity, you will use playdough, cutout pieces of paper, chalk, or some other material to create a model of how that transport occurs. Be sure to include all the terms and concepts in the list below.

Use your model to demonstrate how both water and K^+ ions are transported from the soil into the xylem of the plant and to the leaves of the plant.

Terms

endodermis	symplastic transport
oxidative phosphorylation	ATP
Casparian strip	apoplastic transport
oxygen	epidermis
plasmodesmata	symplast
respiration	lateral transport
stele	apoplast
channels	absorption
root hairs	xylem
ADP	$-\psi$ versus $+\psi$

Use your model to answer the questions.

1. How do water's properties of adhesion and cohesion help maintain the flow of water in the xylem of a plant?

2. a. If water flows from a region of more positive (higher) water potential to a region of more negative (lower) water potential, how does the water potential in the root compare to that in the soil outside the root?

b. How does the water potential in the air compare to that in the leaf of a plant undergoing transpiration?

3. A student uses an $^A U^B$ tube for a series of experiments. Sides A and B of the tube are separated by a membrane that is permeable to water but not to sugar or starch. What results would you expect under the experimental conditions given below?

Explain your answers in terms of osmotic potential, water potential, and the equation

$$\psi = \psi_P + \psi_S$$

(*Hint*: Solute pressure is always negative and a 0.1 *M* solution of any substance has $\psi_S = -0.23$. Therefore, a 0.2 *M* solution would have $\psi_S = -0.23 \times 2 = -0.46$.)

Experiment a: A solution of 10 g of sucrose in 1,000 g of water (the molecular weight of sucrose is 342) is added to side A. An equal volume of pure water is added to side B. What will happen to the concentrations of water and sugar in the two sides over time? Explain.

Experiment b: A solution of 10 g of soluble starch in 1,000 g of water is added to side A. Assume the molecular weight of soluble starch is about 63,000. An equal volume of pure water is added to side B. What will happen to the concentrations of water and starch in the two sides over time? Explain how this compares with the results in Experiment a.

Name_____   Course/Section_____

4. Fertilizer generally contains nitrogen and phosphorus compounds required by plants. The nitrogen is often in the form of nitrates, and the phosphorus is in the form of phosphates. Based on what you know about chemistry and water potential, why would overfertilizing lead to the death of plants?

5. a. One of the most common ways of killing a plant is overwatering. Why does overwatering kill a plant?

    b. If overwatering kills plants, why can you sprout roots from cuttings of stems in water?

6. Xylem cells are dead when functional. Why must phloem cells be alive when functional?

7. What forces bring about:	How are these forces generated in the plant?
a. xylem conduction?	
b. phloem conduction?	

8. Refer to your diagram of the cross-sectional structure of a typical angiosperm leaf from Activity 35.1. Explain how this structure (that is, the type and placement of cells, and so on) is correlated with the activities of the leaf as they relate to photosynthesis, water conservation, and food and water transport.

Name_____   Course/Section_____

36.1 Test Your Understanding

1. Scientists have measured the circumference of trees at 2 A.M. and at 2 P.M. If they collect measurements when the ground has adequate moisture and the days are sunny and dry, they find that the circumference (and therefore the diameter) of the tree trunk is smaller at 2 P.M. than at 2 A.M. From your knowledge of the mechanisms of water transport, suggest the reasons for this decrease in circumference.

2. Outline an experiment that would allow you to determine:

   a. how fast a substance is carried in the xylem,
   b. in what direction the substance flows in the xylem, and
   c. what percentages of solutes are in the xylem at various distances away from the leaves or roots.

3. When researchers have tried to tap into phloem cells during experiments, they find that the disrupted phloem immediately stops functioning. However, aphids can pierce through plant tissues with their mouthparts and locate individual phloem cells. Once inside a phloem cell, the aphids are essentially force-fed phloem sap. If the aphid body is removed from the mouthparts, phloem sap will continue to flow and can be collected (as honeydew).

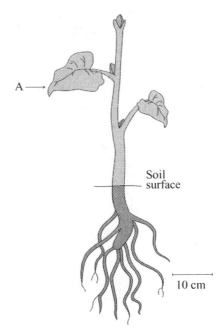

You place two aphids on a plant you are using for radioactive tracer experiments. One is located 5 cm above leaf A and the other is located 5 cm below leaf A. Once the aphids begin feeding, you remove the aphid bodies, allowing you to sample phloem sap from the mouthparts remaining in the plant. You then cover leaf A with a plastic bag and inject $^{14}CO_2$ (radioactive carbon dioxide) into the plastic bag continuously for 10 minutes, during which time leaf A continues to conduct photosynthesis.

You start collecting honeydew from the two aphids' mouthparts from the time the $^{14}CO_2$ is first injected into the bag and every 30 seconds thereafter for a period of one hour.

If you analyze the honeydew samples for ^{14}C, where would you expect to find it? Explain your reasoning.

4. On a separate sheet of paper, outline an experiment that would allow you to determine:
    a. how fast a substance is carried in the phloem,
    b. in what direction the substance flows in the phloem, and
    c. what percentages of solutes are in the phloem at various distances away from the leaves or roots.

Name_____   Course/Section_____

Date_____   Professor/TA_____

 Activity 37.1 What do you need to consider in order to grow plants in space (or anywhere else for that matter)?

Long-range, human-operated space travel and space stations may someday become a reality. Before this can occur, however, we will have to develop sustainable methods of agriculture suitable for use in space. One of the key methods being investigated is hydroponics, growing plants in water supplemented with nutrients.

You are assigned to a team working on the design of plant growth systems for use in a space station.

1. What types of plants would you choose to grow? Explain the reasoning behind your choices.

2. When you set up the growth system:

a. What nutrients would you need to add to the water? List what you would need, and why each would be necessary.	b. What atmosphere would you need to maintain? List what components you would need to maintain in the atmosphere and why each would be necessary.	c. Which of the requirements in parts a and b could be recycled? What could not be recycled? Explain.

Name_____   Course/Section_____

Date_____   Professor/TA_____

Activity 38.1 How can plant reproduction be modified using biotechnology?

Answer questions 1–7. Then in question 8 you will develop a concept map that links the information from all seven answers.

1. Draw a general diagram of the life cycle of a seed plant. Indicate which steps are haploid and which are diploid.

2. Draw and label all parts of a complete flower. Indicate the functions of the major parts.

3. Draw and label a mature ovule. Include the micropyle, integuments, nucellus, synergids, polar nuclei, egg, and antipodals. Indicate the functions of each of these structures.

4. Define microsporogenesis and megasporogenesis. In what part(s) of the flower does each of these processes occur? What is the end product of each process?

5. What is pollination? How does it differ from fertilization?

6. What stages of the life cycle are eliminated or bypassed when plants are cloned naturally? When plants are cloned on the farm or in the laboratory?

7. What does the science of plant biotechnology do that artificial selection and/or cloning practices don't do?

8. Construct a concept map that links all of the information in questions 1–7. Do this on a separate piece of paper.

Name_____   Course/Section_____

Date_____   Professor/TA_____

Activity 39.1 How do gravity and light affect plant growth responses?

Review Chapter 39 of *Biology,* 8th edition, and your answers to Activity 37.1. Then answer the questions.

1. One of the problems associated with growing plants in space is lack of gravity.		
a. How does gravity affect the normal growth of a plant's roots, stems, and other parts? Explain the mechanisms involved.	b. How would a lack of gravity affect normal plant growth?	c. Propose mechanisms to overcome the problems associated with a lack of gravity.

2. Another problem with growing plants in space relates to a plant's light requirements and phototropic responses versus the photoperiods required for the plant to flower and produce fruit.

a. How do phototropism and photoperiodism differ?	
b. What light charac-teristics would you use to maximize plant growth per unit time?	
c. What kind of physical environment would you need to maintain appropriate phototropic responses among plants?	
d. What design modi-fications would you need to make to support plants with different photoperiods—for example, long-day versus short-day plants?	

Activity 39.1

39.1 Test Your Understanding

1. Compared to the wild type, a mutant plant that overproduces cytokinin will probably have ____ lateral branches. Fill in the blank with the best answer and explain your answer.

   a. more
   b. fewer
   c. the same number of

2. Combining the mutation in question 1 above with one that causes the overproduction of auxin would _____ the effect you chose above. Fill in the blank with the best answer and explain your answer.

   a. increase
   b. decrease
   c. not change

3 and 4. A species of long-day plant will flower when the days are over 14 hours long. Which of the following light cycles would cause this plant species to flower?

Answer **True** if the cycle causes flowering and **False** if it does not. Explain your answers.

T/F   3.   Continuous white light day and night, interrupted every 4 hours by flashes of red light.

T/F   4.   A daily cycle of 9 hours of light followed by three periods of 4 hours and 50 minutes of uninterrupted dark, and 10 minutes each of red light and white light.

Name_____   Course/Section_____

Date_____   Professor/TA_____

 Activity 40.1 How does an organism's structure help it maintain homeostasis?

1. To remain alive, an organism must be able to maintain homeostasis of its internal environment relative to the external environment. What behaviors, structure(s), or system(s) are of primary importance in maintaining homeostasis in the following situations in amoeba versus mammal?

Situation	Amoeba	Mammal
a. Change in environmental: pH temperature	*Behavioral:* *Structural and physiological:*	*Behavioral:* *Structural and physiological:*
b. Reception of stimuli, for example: light touch		
c. Response to stimuli		

2. Cells must be bathed continuously in an aqueous medium to take in oxygen and nutrients and get rid of waste products via diffusion. Diffusion is efficient over only short distances. In fact, diffusion is efficient only for a distance of about three cell diameters maximum (approx. 200 to 300 μm). Note the following times required to diffuse specific distances:

Diffusion Distance (μm)	Time Required for Diffusion
1	0.5 msec
10	50 msec
100	5,000 msec (5 sec)
1,000 (1 mm)	500,000 msec (8.3 min)
10,000 (1 cm)	50,000,000 msec (14 hr)

a. Graph the data in the table and calculate the mathematical relationship between the increase in distance and the corresponding increase in time required to diffuse that distance. You can do this by hand or you can use a software program with graphing capability. Include your graph and the results of your calculations below.

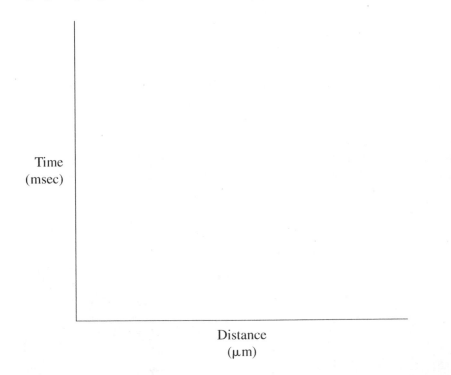

Time
(msec)

Distance
(μm)

b. How much time would be required for oxygen to diffuse 5 μm? 200 μm?

3. What happens to the surface-area-to-volume (SA/V) ratio of a three-dimensional object (such as a cell) as its linear dimension increases? For example, how does the SA/V ratio of a sphere or cube change as the linear dimensions increase? (Formulas for a sphere: surface area $= 4r^2$; volume $= {}^4/_3 r^3$.) (Also see Activity 7.1.)

40.1 Test Your Understanding

Propose the effect(s) the physical properties of diffusion are likely to have on the structure and function of epithelia and epithelial cells and digestive and circulatory systems as animals become larger and larger. In your answer, consider how SA/V ratios change as organisms become larger and the effect(s) this change is likely to impose on the structure and function of organisms.

Name_____   Course/Section_____

Date_____   Professor/TA_____

Activity 41.1 How are form and function related in the digestive system?

1. What is the overall function of digestion (as a whole)?

2.
a. How do bacteria eat?	b. How do amoebas eat?
How are their eating styles similar to what we see in humans? How are they different?			
c. Similarities with bacteria	d. Differences from bacteria	e. Similarities with amoebas	f. Differences from amoebas

3. Label the parts of the mammalian digestive tract in the diagram, and state the major function(s) of each part.

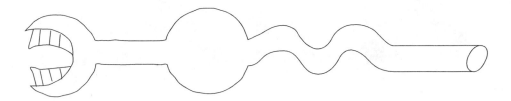

How does the structure of each part reflect its function?

a. Mouth and teeth	b. Esophagus	c. Stomach	d. Small intestine	e. Large intestine

4. Make a larger copy of the diagram in question 3. Use this larger diagram to model or trace what happens to a food particle from the time it enters the mouth until its indigestible remains are egested, or eliminated. Include all of the following terms in your discussion. Also, note on the diagram the function of each and where it is found in the digestive system.

carbohydrate

fat

nucleic acid

protein

cardiac orifice

epiglottis

pyloric sphincter

pharynx

pepsinogen

lipase

pepsin

salivary amylase

bile

dipeptidase

saliva

amylase

pancreas

nuclease

gastric acid (HCl)

bolus

microvilli

lacteals

capillaries

feces

intestinal bacteria

hepatic portal vein

fiber or roughage

Using the understanding of the structure and function of the digestive system you gained from the model in question 4, answer the questions.

5. The mammalian digestive tract has been called an extension of the outside world that you enclose in your body.

   a. What does this statement mean? Consider what would happen if you swallowed a marble. Is the marble ever "inside your body?"

   b. At what point in the digestive process is food officially inside the body?

   c. How is mammalian digestion more efficient than the type of digestion seen in bacteria?

6. Digestion in humans and many other animals is both physical and chemical. Among the chief chemical agents of digestion are the digestive enzymes. What do enzymes do to food?

7. Have you ever heard the old adage: "Be sure to chew your food 20 times before swallowing?"

   a. What, if any, effect would this chewing have on how well the digestive system functions? Keep in mind that enzymes work only on the surfaces of food particles. Explain.

   b. How does the function of the teeth complement the function of one of the digestive chemicals in the stomach? Be sure to name the specific chemical in your answer.

8. Although enormous quantities of various enzymes are added to the contents of the duodenum of the small intestine, no traces of enzymatic activity are left in the intestinal contents when they pass into the large intestine (colon). Why? What happens to the enzymes?

9. Most of the blood that leaves the digestive tract of a human is collected into a series of veins that merge to form the hepatic portal vein. The hepatic portal vein carries blood to the liver, where the hepatic portal vein divides again into a system of venules and then capillaries. The liver capillaries drain into the hepatic vein, which carries blood to the vena cava. The vena cava carries blood from the body to the right atrium of the heart. Some of the products of digestion enter a different system of transport, the lacteal system. The lacteal system bypasses the liver and carries its contents directly to the right atrium of the heart.

   a. Which products of digestion are carried in the blood to the liver?

   b. Which products of digestion are carried via the lacteal system?

   c. During the first hour after a heavy meal, how does the concentration of glucose in the blood going from the small intestine to the liver compare to the concentration entering the right atrium of the heart?

   d. Similarly, how does the concentration of amino acids compare?

   e. How does the concentration of fat leaving the small intestine compare to the concentration in the right atrium?

10. How does an herbivore such as a cow extract the glucose from the cellulose in its diet? What characteristics of the structure and function of the digestive tract of a ruminant suit it for this function?

41.1 Test Your Understanding

1. A peanut butter and jelly sandwich contains a variety of carbohydrates, fats, and proteins. Complete the graph below to indicate the relative percentages of carbohydrate, fat, and protein that remain in this ingested food as it progresses from your mouth through your digestive tract. Explain your reasoning.

Key: Protein = xxxxx; Carbohydrate = ---------; Fat = oooooooo

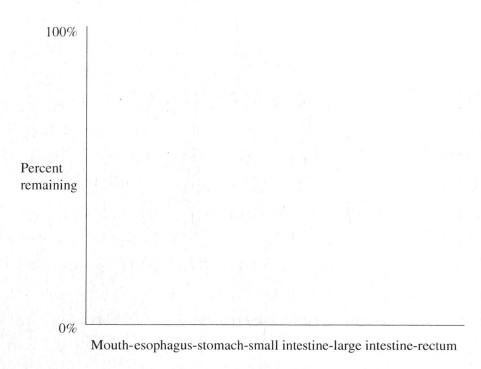

Mouth-esophagus-stomach-small intestine-large intestine-rectum

2. A good rough generalization is that the more meat in the diet of a species of animal, the shorter its intestine. In comparison, herbivores have long intestines (length always being relative to total body length). How can this be explained?

3. What would happen to the normal function of the digestive tract if part of one of the following organs was removed or greatly reduced in size (for example, as the result of surgery following an accident)? How would a person's eating habits need to change to accommodate the reduction in size?

    a. Stomach

    b. Small intestine

    c. Large intestine

4. A friend of yours has come up with a new idea for people who don't have time to eat. He has developed a high-powered Blenderizer that breaks food up into very small particles. He has tested his product on amoeba and *Paramecia*. Both of these single-celled organisms can pick up these small particles and incorporate them into food vacuoles. Your friend is hoping to market the Blenderizer to busy people who could blenderize their food and use an IV bag and tube to run the food directly into their blood system. Because people could hide the system under their clothes, they could essentially eat any time without it interfering with meetings and other activities. Your friend comes to you for advice and possible financial support to get his idea off the ground. What would you say to him?

Name_____   Course/Section_____

Date_____   Professor/TA_____

Activity 42.1 How is mammalian heart structure related to function?

1. Diagram and describe the path a red blood cell takes from a capillary in your big toe to your heart and back to your big toe. At each point in the pathway, indicate whether the red blood cell is most likely picking up or losing oxygen. Indicate also the relative blood pressure that part of the path is likely to have. Include all of the following terms in your diagram.

aorta

venules

veins

arteries

arterioles

capillaries

right atrium

left atrium

right ventricle

left ventricle

coronary arteries

heart

internal organs (for example, lungs, digestive tract)

skeletal muscle

2. The degree of musculature differs in these chambers of the heart: atria, right ventricle, left ventricle.

a. Draw the heart and its chambers, including differences in musculature.	b. Explain why the differences in musculature might exist by explaining the normal functions of each chamber.	c. Include below the functions of the SA node (pacemaker), the AV node, and the AV and semilunar valves.

Name_____   Course/Section_____

42.1 Test Your Understanding

1. While in utero, the wall between the two atria of the human fetal heart is not complete. An opening, the foramen ovale, allows blood from the two atria to mix. Normally, at birth, this hole seals over and the two atria are separated from each other. What would be the consequences to the infant if this hole did not seal over at birth?

2. One of the most common congenital defects of the cardiovascular system is called "transposition of the great arteries." In infants who have this defect, the pulmonary artery exits from the heart where the aorta should and the aorta exits where the pulmonary artery should. All other circulatory connections are normal.

   a. Diagram and describe the circulation of blood in an infant who has this genetic defect.

   b. What type of treatment would such an infant need?

Activity 42.2 How do we breathe, and why do we breathe?

This activity is designed to give you an understanding of how and why we breathe and some of the mechanisms that control the rates of breathing.

How do we breathe? Explain the mechanics of breathing. To do this, diagram (or model) and explain the general movement of oxygen and carbon dioxide into and out of the lungs, lung alveoli, blood, and tissues. Be sure to include all of the following terms in your diagram.

diaphragm

diffusion

CO_2 removal

O_2 demand

lungs

alveoli

expired air

inspired air

oxygen

carbon dioxide

brain

P_{CO_2} sensors

medulla oblongata

lung tissue cells

skeletal muscle cells

carbohydrate $+ O_2 \rightarrow CO_2 + H_2O + $ energy

$CO_2 + H_2O \leftrightarrow H_2CO_3 \leftrightarrow HCO_3^- + H^+$

Use your diagram or model to answer the questions.

1. Explain briefly why we breathe. Or, more specifically, where is what we breathe in used in the body and where is what we breathe out produced in the body?

2. Oxygen is transported in the blood in reversible combination with hemoglobin.

   a. How does the level of carbon dioxide in the blood contribute to blood pH? (Write the chemical reaction.)

   b. How does the level of CO_2 in the blood affect the affinity of hemoglobin for oxygen? That is, how does the concentration of CO_2 in the blood change its ability to carry oxygen (the effect of the Bohr shift)? What is the Bohr shift?

   c. Is the Bohr shift more likely to occur after exercise or after hyperventilation (rapid breathing with no exercise involved)? Explain.

3. Compare these oxygen dissociation curves for hemoglobin.

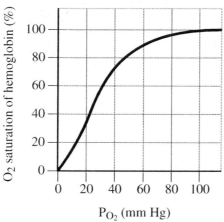

Normal adult

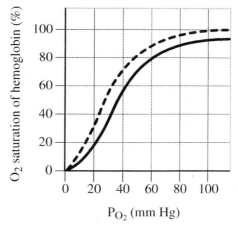

Reduced pH conditions

Note: The solid line indicates the curve for reduced pH conditions.

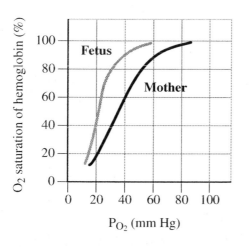

Fetal vs. Adult

a. How do the curves differ?
   i. Normal versus reduced pH?
   ii. Fetus versus adult?

b. What would happen if fetal and adult hemoglobin had the same affinities for oxygen?

4. If they don't get what they want, some small children threaten to "hold their breath until they die." Children who are very strong-willed may turn blue, but they will not die because they will pass out and the autonomic breathing response will take over. What types of control are involved in holding your breath? Why (for what physiological reason) will you pass out if you hold your breath too long?

Name_____   Course/Section_____

Date_____   Professor/TA_____

Activity 42.3 How are heart and lung structure and function related to metabolic rate?

1. Describe the simplest type of closed circulatory system. Include the function of each part of the system and how the structure of each part is related to its function.

2. If you want to increase the rate of metabolism in an organism, you need to increase: (a) the rate of food and oxygen delivery to the cells (or body) and (b) the rate of removal of waste products from the cells (or body). In the evolution of the vertebrates, what modifications of the circulatory system and/or circulatory function have occurred to increase metabolic rate?

3. Both reptiles and mammals obtain their energy via aerobic respiration:

$$C_6H_{12}O_6 + 6O_2 \rightarrow 6CO_2 + 6H_2O + 36ATP$$

Both also need to be able to maintain internal homeostasis in order to survive.

a. Given this, would you expect the lungs of the lizard (an ectotherm) to have more or fewer alveoli per unit area than the lungs of the rat (an endotherm)? Assume the overall *SA/V* ratios of the rat and lizard are about the same. Explain the physiological reasoning behind your answer.

b. On average, would you expect the heart rate of the lizard to be higher or lower than the heart rate of the rat? Explain the physiological reasoning behind your answer.

4. Based on your knowledge of *SA/V* ratios, metabolic rates, and cardiopulmonary physiology, would you expect the following to be higher or lower in a human infant as compared to an adult? Explain the physiological reasoning behind your answer(s).

a. body temperature

b. heart rate

c. oxygen consumption rate/unit mass of body

42.3 Test Your Understanding

1 to 3. At rest, an average human's mean arterial blood pressure (MABP) is 120/80. As a person increases activity, however, a variety of modifications must be made to support the increased activity.

For example, assume you are late for an exam and run four blocks to get there on time. Which of the following would occur during your four-block run? Explain your answers.

T/F   1.   As you run, your muscles will require more oxygen per unit time than when you walk.

T/F   2.   The percentage of carbon dioxide in your blood will increase and your blood pH will increase.

T/F   3.   Your MABP will increase because both the heart rate and stroke volume will increase.

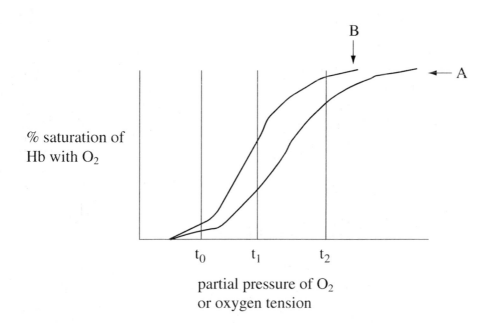

Refer to the preceding graph of dissociation curves for two mammalian hemoglobins to answer questions 4 to 7. Explain your answers.

T/F   4.   At oxygen tension t_0 both hemoglobins have released most of their oxygen.

T/F   5.   At oxygen tension t_1 hemoglobin B is carrying less oxygen than hemoglobin A.

T/F   6.   If comparison of curves A and B illustrates the Bohr effect in a single species, then curve B represents the dissociation at the higher concentration of carbon dioxide (lower pH).

7.   In addition to carbon dioxide, contracting muscles cells produce a large amount of heat energy. It has been demonstrated that an increase in temperature, independent of carbon dioxide levels, encourages the blood to give up its oxygen to exercising tissues. Given this information, would you expect a or b below to be true? Explain your answer.

a.   Curve A of the graph would be the lower-temperature curve and B would be the higher-temperature curve.

b.   Curve B of the graph would be the lower-temperature curve and A would be the higher-temperature curve.

Name_____   Course/Section_____

Date_____   Professor/TA_____

Activity 43.1 How does the immune system keep the body free of pathogens?

Draw a Rube Goldberg cartoon-type diagram or develop a dynamic (claymation-type) model to demonstrate how the components of the immune system interact to rid the body of a pathogen—for example, a bacterial cell or a viral particle. Be sure to explain the function of each "actor" in the system. Your diagram or model should include all the terms below.

Here is an example of a Rube Goldberg–type drawing:

Terms

bacterium or virus particle	memory B cell	epitope
helper T cell receptor	memory T cell	thymus
helper T cell	plasma cell	bone marrow
cytotoxic T cell	interleukins (or cytokines)	hypothalamus
active cytotoxic T cell	CD4 protein	fever
macrophage	MHC molecules	clonal expansion
B cell	antibody	self versus nonself
memory helper T cell	antigen	

After you have completed your model or diagram, use what you have learned to answer the questions on the next page.

1. What are pathogens? Why do we need to prevent them from colonizing our bodies? If pathogens do manage to colonize, what effects can they have?

2. What general defense mechanisms does the body use to help prevent colonization by pathogens? For example, what general defense mechanisms are involved in local inflammatory responses?

3. In specific immunity, how do B cell responses differ from T cell responses?

B cell responses	T cell responses

4. If about 10^5 genes are available in the human genome to produce proteins, how can we produce more than 10×10^6 different kinds of Ab receptors (proteins) on B cells?

5. How does HIV affect the immune system?

Name_____   Course/Section_____

Date_____   Professor/TA_____

Activity 44.1 What is nitrogenous waste, and how is it removed from the body?

In the space below, draw a longitudinal section of a mammalian kidney. Next to this, draw a blowup of a nephron (including Bowman's capsule and the loop of Henle) and its associated collecting duct. Be sure to include the afferent arteriole, glomerulus, and efferent arteriole that are associated with this nephron.

You may do your drawing in chalk on a tabletop or blackboard if they are available.

Use your drawing and your understanding of the operation of the kidney to answer the questions.

1. Define excretion, and indicate how it differs from elimination.

2. The removal of nitrogenous wastes (excess nitrogen) is a special problem in most animals.

   a. Where does the nitrogenous waste come from?

   b. What is it about the chemistry of nitrogen that makes it difficult for most animals to deal with?

Activity 44.1

Name_____   Course/Section_____

3. Work through parts a and b, and then use the information you gather there to answer the question in part c.

   a. Describe the composition of the newly filtered solution that enters Bowman's capsule. Then compare it to the composition of the blood entering and leaving the glomerulus.

   b. Starting with the solution that escapes into Bowman's capsule from the glomerulus, describe the changes that occur in its composition as it moves through each of these regions:

      i.   Proximal convoluted tubule

      ii. Loop of Henle

      iii. Distal convoluted tubule

      iv. Collecting duct

      v.   Urinary bladder

   c. Now explain how the general function of the kidney enables it to "remove" (a better expression would be "let out") from the body a wide variety of unfamiliar substances (drugs, inorganic molecules, or ions of many kinds) that the body has never encountered before. After answering this, explain why "let out" from the body is a better expression than "remove."

Activity 44.1   255
Copyright © Pearson Education, Inc., Publishing as Benjamin Cummings

4. It is useful to consider the excretory system (along with the digestive and gas-exchange systems) as primarily involved in bulk exchange with the external environment. The excretory system could also be interpreted as a specialized part of the external surface of the organism, which in its own way encloses and modifies part of the environment. Describe how this is true for the human kidney. For example:

   a. Where in the kidney does the organism end and the environment begin?

   b. Are changes in the glomerular filtrate changes in the organism, changes in the environment, or both?

   c. What do your answers in parts a and b indicate about the possible evolutionary origins of the kidney?

5. a. What is the difference between hydrostatic pressure and osmotic pressure?

   b. Where in the human excretory system is hydrostatic pressure responsible for moving water across a membrane or layer of cells?

   c. Where in the excretory system is osmotic pressure responsible for such movement?

Name_____   Course/Section_____

1. a. You examine the kidney structure and function of a two species of mice, one from the desert and another from a meadow or grassland. What differences would you expect to find? Use drawings of the two systems and their functions to explain your reasoning.

    b. Would you expect the excretory systems of organisms that live in the sea to resemble more closely those of animals that live in deserts or animals that live in fresh water? Explain.

2. Many medicines are taken orally—that is, swallowed and absorbed from the digestive system. Aspirin and many antibiotics are examples. Even though these drugs may be very different chemically, the instructions for taking them often say "Repeat dose every 4 to 6 hours." What is the reasoning behind this dose rate? In other words, why do you need to take the medicine every 4 to 6 hours?

Indicate whether the following are true or false. Explain your reasoning.
Questions 3 to 5. In a normal, healthy human, urine:

T/F   3.   includes nitrogenous waste products formed following deamination of amino acids.

T/F   4.   functions to rid mammals of excess glucose.

T/F   5.   production involves a countercurrent principle of exchange.

Questions 6 to 8. Injection of an individual with antidiuretic hormone (ADH) would cause:

T/F   6.   increased urine volume.

T/F   7.   decreased filtration at the glomerulus.

T/F   8.   an increase in permeability of the collecting ducts.

Name_____   Course/Section_____

Date_____   Professor/TA_____

Activity 45.1 How do hormones regulate cell functions?

This activity is designed to help you understand how hormones act within cells to produce a response.

Building the Model

Working in groups of three or four, construct a dynamic (claymation-type) model of hormone action in cells. You may use the materials provided in class or devise your own.

Step 1. Use chalk on a tabletop or blackboard to draw a eukaryotic cell. Your cell should be at least 18 inches in diameter. Be sure your drawing includes the cell membrane, the nuclear membrane, and the DNA inside the nucleus.

Assume that your eukaryotic cell responds to two different hormones:

- Hormone 1 is a protein-derived hormone. The cell responds to hormone 1 by increasing production of substance X.
- Hormone 2 is a cholesterol-derived hormone. The cell responds to hormone 2 by decreasing production of substance Y.

Step 2. Answer the following questions.

1. What structures or components do you need to add to your model to allow hormone 1 to react and increase production of substance X?

2. What structures or components do you need to add to your model to allow hormone 2 to decrease production of substance Y?

Now use playdough or cutout pieces of paper to make your hormones, cell membrane proteins, and any other proteins you need. Indicate their placement on the membrane or cell. Include a key for your model that indicates how different hormones and proteins are designated.

Step 3. Using claymation, demonstrate how each of the two hormones is likely to produce its response.

Use the understanding you gained from your model to answer the questions.

3. In medical applications, the type of hormone dictates the mode of administration—for example, oral versus injection, and so on.

   a. How would you need to administer hormone 1 to an organism deficient in this hormone?

   b. How would you need to administer hormone 2?

4. Hormones often act in an antagonistic fashion. That is, one hormone will initiate a certain response while another inhibits that response. Illustrate this process using insulin and glucagon as examples of antagonistic hormones.

45.1 Test Your Understanding

1. Assume you are trying to characterize one of the hormones that is involved in the neuroendocrine regulation of milk production. You make extracts of blood and mammary tissues of normal, lactating (milk-producing) animals and assay these extracts by injecting them into animals (each of whose hypothalamus has been surgically removed) and look for restoration of milk production. Indicate whether each of the following findings is consistent with the hormone being a steroid, a peptide, or either type. Explain your reasoning.

   a. The hormone is found in blood.

   b. It is found in the cytoplasm of mammary tissue cells.

   c. It is found associated with receptors.

   d. It is found associated with protein complexes that contain G protein.

   e. It is found in nuclear extracts of cells.

2 to 4. Glucagon and insulin are hormones that act homeostatically to maintain glucose levels in the body. Are the following statements **True** or **False** concerning insulin and glucagons? Explain your reasoning.

T/F   2.   Their production and release is stimulated by trophic hormones from the anterior pituitary.

T/F   3.   They are examples of antagonistic hormones.

T/F   4.   They act by stimulating storage or release of glucose from cells.

Name_____   Course/Section_____

Date_____   Professor/TA_____

 Activity 46.1 How does the production of male and female gametes differ in human males and females?

This activity is designed to help you understand how gamete production is controlled in mammals and particularly in humans.

In human males and females, the production of gametes and the hormones estrogen, progesterone, and testosterone is ultimately controlled by actions of the hypothalamus.

Using all the terms below, diagram the control of gamete and sex hormone production first in a human male and then in a human female. Be sure to explain the role(s) of each term in your diagram.

hypothalamus   progesterone

anterior pituitary   estrogen or testosterone

LH   secondary sex characteristics

FSH   primary sex characteristics

ovary or testes   negative feedback

follicle or seminiferous tubule   egg and polar bodies or sperm

corpus luteum or Leydig cells

Human male:

Human female:

Use your diagrams to answer the questions.

1. In both males and females, the hypothalamus produces GnRH, which stimulates the pitutitary to release LH and FSH. Fill in the chart.

Hormone	a. In males causes:	b. In females causes:
LH		
FSH		

2. In both males and females, the testes or ovaries produce additional hormones.

	a. Males produce:	b. Females produce:
Hormones		
Function of these hormones produced in the gonads		

3. Most birth control methods are designed to keep the egg and sperm from uniting to form a zygote. Many birth control pills or patches used by human females contain a combination of estrogen and progesterone. How do they keep sperm from uniting with egg? Explain the mechanism.

4. Efforts to make a male contraceptive pill (analogous to the pills used by females) have not been very successful. Given what you know about the similarities and differences in male and female gamete production, propose why this might be the case.

5. Fertilization generally occurs in the upper third of the oviduct, and development of the fetus occurs in the uterus. In some relatively rare cases, however, developing embryos have attached to the outside of the uterus and developed there for the full nine months of pregnancy.

   a. Given the anatomy of the female reproductive system, can you explain how this could happen?

   b. What modifications of normal birthing procedures (if any) would have to be made in such cases?

46.1 Test Your Understanding

1 to 3. In an experiment, an adult rat's testes, including the vascular connections, were transplanted to the wall of the abdomen. Connections of the testes to the reproductive tract were cut/severed.

Following recovery, which of the following would be true for this rat? Explain your answers.

T/F   1.   The rat would have lowered sexual activity due to loss of testosterone.

T/F   2.   The rat would have normal sexual activity but be unable to produce any ejaculate.

T/F   3.   The rat would have normal sexual activity but have no sperm in the ejaculate.

4.   A girl begins to develop breasts and pubic hair at the age of four. Given these symptoms, her physician orders a CT scan (imaging procedure) to look for an endocrine tumor. Which organ would he most likely **not** investigate as the cause? Explain your reasoning.
   a. hypothalamus
   b. pituitary
   c. ovary
   d. uterus

5 to 8. Assume that women can be vaccinated against the following hormones. Each vaccine is designed to completely neutralize the target hormone's activity. Which vaccine(s) would prevent pregnancy? Explain your answers.

A = Would prevent pregnancy; B = would not prevent pregnancy

5.   A vaccine against LH

6.   A vaccine against CG

7.   A vaccine against estrogen

8.   A vaccine against prolactin

Name_____   Course/Section_____

Date_____   Professor/TA_____

Activity 47.1 What common events occur in the early development of animals?

The early stages in the development of all vertebrates (and many other animals) include zygote formation, cleavage, blastula formation, gastrula formation, and organogenesis (for example, neurulation). Among the major aspects of development are cell adhesion/recognition, cell growth (in number and/or size), cell induction, and cell determination.

1. What key events occur at each stage of development?

Developmental stage	What occurs during this stage?	What is the influence or effect on the subsequent development of the embryo?
a. Cleavage		
b. Blastula formation		
c. Gastrula formation		
d. Organogenesis (for example, neural tube formation)		

2. Many animal species share these similarities in early development. Yet, the stage at which the individual cells of the embryo lose their totipotency can vary considerably among these species.

   a. What does it mean to say that a cell is totipotent?

   b. What factors can affect the point at which a cell loses its totipotency—that is, when its fate becomes determined?

Name_____   Course/Section_____

47.1 Test Your Understanding

1 and 2. Choose the graph that best fits each situation.

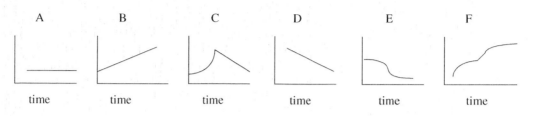

1. Which graph best describes the change in size of individual cells of a vertebrate embryo from the time of zygote formation to the end of cleavage? Explain your answer.

2. Which graph best describes the change in the number of cells in the embryo from fertilization to gastrulation? Explain your answer.

3 to 6. While the mechanics of gastrulation may differ among various organisms, the overall objectives and problems are the same. Which of the following are accomplished by the end of gastrulation?

T/F 3. formation of the three germ layers—ectoderm, mesoderm, and endoderm

T/F 4. establishment of the embryonic axis (anterior to posterior)

T/F 5. determination of the fates of the individual cells of the gastrula

T/F 6. formation of the neural tube

7. In a now classic experiment, Spemann and Mangold took cells from the dorsal lip of the blastopore from one frog embryo (the donor) and transplanted them to an area opposite the dorsal lip in another frog embryo (the host). As a result, the host embryo developed two heads, one at the site of the "new" or transplanted dorsal lip and the other at the site of the host's original dorsal lip. A thorough examination indicated that *the second head was formed from host cells*.

Which of the following developmental cues or mechanisms was most likely the trigger (or cause) for the generation of the second head? Explain your answer.

    a. cytoplasmic determinants that are unequally distributed in the host embryo
    b. gradients set up in the egg that gave rise to the host embryo
    c. *Hox* genes (or homeobox genes) present in the embryo receiving the transplant
    d. cell-cell signalling between the transplanted dorsal lip and the host embryo
    e. all of the above

Name_____   Course/Section_____

Date_____   Professor/TA_____

? **Activity 48.1 How do ion concentrations affect neuron function?**

Much of our understanding of neuron function was based on studies of the squid giant axon. Squid move through the water by contracting the muscles of the mantle. This compresses water inside the mantle that is forced out the siphon. Squid can change the direction of movement by directing the flow from the siphon either forward or backward (relative to the head or anterior end).

1. In which direction would the squid move if water flow from the siphon was directed toward the head end of the organism?

2. When squid are startled or in danger, they can simultaneously contract all muscles of the mantle to jet water forcefully out of the siphon and escape rapidly. Assume the mantle of the squid is 30 cm in length (about 12 inches). The brain sends a signal to major nerve ganglia in the mantle, which relay the signals to axons innervating the mantle muscles. For all muscles of the mantle to contract simultaneously, all nerve signals sent along these axons must reach all parts of the mantle at the same time.

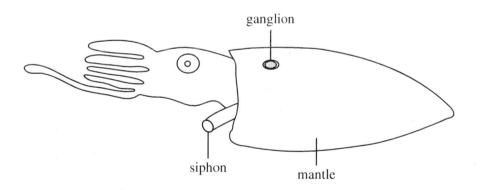

    a. On the diagram of the squid above, draw and number three neurons. Assume all are simultaneously stimulated by a single motor neuron from the brain. Neuron 1 innervates the mantle muscles nearest the brain. Neuron 2 innervates the muscles in the midregion of the mantle. Neuron 3 innervates the muscles at the tail end of the mantle.

    b. In invertebrates like the squid, how must the nervous system be structured to allow both the muscles nearest the brain and those farthest from it to contract simultaneously? In your drawing, indicate any differences in the size or structure of the three neurons that would be required. Explain your reasoning.

Activity 48.1   271

3. Researchers discovered that they could remove the giant axons of the squid. With some skill, the axons could be maintained outside the body if held in ion solutions of the same concentrations as the squid's extracellular fluids. If stimulated with a microelectrode, these isolated axons would generate action potentials. By recording the potential difference between an electrode in the axon versus one in the fluid bathing the axon, the scientists could also record the potential difference at rest and any change in potential difference that occurred as a nerve impulse, or action potential, was generated.

The ion concentrations the researchers recorded for the squid giant axon are listed in the table below.

Ion concentrations (mM) for intra- and extracellular fluids of the squid axon		
	Intracellular (mM)	Extracellular (mM)
Potassium (K^+)	400	20
Sodium (Na^+)	50	440
Chloride (Cl^-)	40 – 150	560
Calcium (Ca^{++})	1×10^{-4}	10

Some of these ion concentrations for the human are listed in the table below.

Ion concentrations (mM) for intra- and extracellular fluids of the human axon		
	Intracellular (mM)	Extracellular (mM)
Potassium (K^+)	140	5
Sodium (Na^+)	15	150

Using the Nernst equation, your textbook calculates that in humans the equilibrium potential for potassium is -90 mV and that for sodium is $+62$ mV.

What are the equilibrium potentials for potassium and sodium for the squid?

The Nernst Equation $E_{ion} = \dfrac{62mV}{z} \log \dfrac{[ion]_{out}}{[ion]_{in}}$ $z = $ ion valance (e.g., $Na = +1, Cl = -1$)

$E_K =$

$E_{Na} =$

4. In a typical neuron, which ion has the greatest influence on the membrane potential at rest? In other words, flux of which ion contributes the most to the resting membrane potential? Explain why this occurs.

5. In calculating the resting and action potentials of the axon, we generally don't worry about the concentration differences of calcium or chloride ions. Explain.

6. What effects would the following changes in extracellular K^+ concentration be likely to have on the resting membrane potential of neurons in a human?

    a. Change extracellular K from 5 mM to 2 mM

       $E_{K^+} =$

    b. Change extracellular K from 5 mM to 10 mM

       $E_{K^+} =$

Name_____   Course/Section_____

Date_____   Professor/TA_____

Activity 48.2 How do neurons function to transmit information?

Working in groups of three or four, construct a dynamic (claymation-type) model of the transmission of an action potential along a neuron and then across a synapse to generate an action potential in a postsynaptic neuron.

When developing and explaining your model, be sure to include definitions or descriptions of the following terms and structures.

Neurons or Parts of Neurons	Ions	Gates
dendrite	K^+	voltage-gated ion channels
axon	Na^+	Na^+ gates or channels
cell body	negative organic ions	K^+ gates or channels
synaptic vesicles	Ca^{++}	Ca^{++} gates or channels
presynaptic neuron		
postsynaptic neuron		

Building the Model

- Using chalk on a tabletop or a marker on a large sheet of paper, draw the membranes of two neurons and the synaptic region between them. Each neuron should each be at least 4 inches across and a foot or more in length.

- Identify the axon, cell body, and dendrite(s) on your drawing.

- Make the ions and gates from playdough or cutout pieces of paper. Indicate the placement of gates in the membranes and ions inside the membrane versus outside the membranes.

- Include a key for your model that indicates how ions and gates are differentiated from each other. You may use color coding for the ions and gates.

- Start the model by "initiating" an action potential in the axon hillock.

- Indicate how this action potential is propagated along the axon and how it can lead to production of an action potential in the postsynaptic neuron.

- When you have completed your model, explain it to another student or to your instructor.

Use your understanding of how action potentials are generated and propagated to answer the questions.

1. All cells maintain an ionic (and therefore electrical) potential difference across their membranes. In most cells, this potential difference is between -50 and -100 mV. That is, the inside of the cells is more negative than the outside by 50 to 100 mV. Although all cells in the body maintain this potential difference across their membranes, only certain cells (for example, neurons) are capable of generating action potentials.

   a. How is this potential difference across the cell membrane generated?

   b. What characteristics of membranes allow cells to concentrate or exclude ions?

   c. What is it about neurons (nerve cells) that makes their properties different from those of other cells? In other words, what enables nerve cells to produce action potentials?

   d. How is an action potential started and propagated?

   e. Is any direct or indirect energy input required to generate an action potential? If so, when and where is the energy used?

f. What happens in time and space (along the axon) once an action potential begins?

g. What factors ultimately limit the ability of the nervous system to respond (that is, to continue to generate impulses)?

2. If an axon is stimulated in the middle of its length, nervous signals (action potentials) will move out from the point of stimulus in both directions. Normally, however, nerve signals move in only one direction along neurons. Explain.

3. Whether or not an action potential is generated in a postsynaptic neuron depends on a number of factors. What are they? What ultimately determines whether or not an action potential is generated in the postsynaptic neuron?

4. Diffusion is efficient over only very short distances. In fact, as you can see in this table, diffusion is efficient only for distances of about 1 to 100 μm.

Diffusion distance (μm)	Time required for diffusion
1	0.5 msec
10	50 msec
100	5 sec
1,000 (1 mm)	8.3 min

a. How wide is a synapse?

b. If a synapse were two times as wide, what effect would it have on the transmission of nerve signals from one neuron to the next? How would this change affect the response time of an organism?

5. If you examine neuron transmission within an organism, you discover that every action potential generated is stereotyped; for example, every action potential reaches the same maximum height and the same minimum height. In addition, the generation of action potentials is an all-or-none phenomenon. That is, once the potential difference across the membrane reaches threshold, an action potential will be generated. Given this, how does the nervous system signal differences in intensity of signal?

Name_____   Course/Section_____

Date_____   Professor/TA_____

Activity 48.3 What would happen if you modified a particular aspect of neuron function?

In the following questions, test your understanding of the various parts of the nervous system by asking yourself what would happen if a certain part was damaged. What would the system still be able to do? What would it be unable to do?

1. Some nerve gases and insect poisons work by destroying acetylcholine esterase. Acetylcholine esterase is normally present in acetylcholine synapses and acts to degrade acetylcholine. What is likely to happen to nervous transmission in insects exposed to this type of insect poison?

2. The pufferfish (fugu) contains the poison tetrodotoxin. Some shellfish produce a paralytic poison called saxotoxin. Both of these poisons block the Na^+ channels in neurons. What specific effects could these toxins have on neuron function?

3. A type of spider (the funnel-web spider) produces a toxin that blocks the Ca^+ channels.

   a. Can a neuron exposed to this toxin fire an action potential? Explain.

   b. Can a neuron transmit a signal across the synapse using neurotransmitters? Explain.

4. You isolate a section of a squid giant axon and arrange an experiment so that you can change the solution bathing the axon. You insert an electrode into the axon and place another electrode outside the cell so that you can measure the potential across the cell membrane. With the axon bathed in normal extracellular fluid, you observe a resting potential of -70 mV and action potentials, when stimulated, that reach $+55$ mV.

	mM concentration of each ion			
	Normal concentrations		Experimental concentration in (a)	
Ion	Inside neuron	Outside neuron	Inside neuron	Outside neuron
Na^+	50	440	50	440
K^+	400	20	400	40

a. You change the solution bathing the neuron by increasing the K^+ concentration to 40 mM. What effect will this have on the neuron? For example, will it depolarize the membrane and make it easier to start an action potential? Will it hyperpolarize the membrane and make it more resistant to starting an action potential? Or will it have no effect? Explain your answer.

b. What would happen if, instead of adding more K^+ to the outside, you added more Na^+ to the fluid bathing the neuron? Explain.

Name_____   Course/Section_____

Date_____   Professor/TA_____

Activity 49.1 How is our nervous system organized?

Neurons are single cells composed of processes called dendrites, which carry information to the neuron's cell body, and an axon, which carries information away from the cell body. In simple terms, nerves are bundles of dendrites, or axons, or mixed axons and dendrites. Ganglia are collections of nerve cell bodies, and brains are collections of ganglia. Processing and integration of the information occurs as a result of interaction within and between ganglia.

1. Responding to stimuli requires at a minimum:
    a. reception of a stimulus, which triggers
    b. an action potential in a sensory neuron, which synapses with
    c. a motor neuron, which causes a response in
    d. a muscle or gland

To coordinate activity an interneuron is often found between the sensory and motor neurons. The interneuron can function to send information to the central nervous system and as a result lead to a response by the whole organism.

In the space below, draw this minimum system. Be sure to name each neuron and label each of its parts.

2. The blink response is one type of simple reflex response that occurs, for example, in response to water splashing up out of the sink or to a snowball aimed at your face. When these events happen, you blink; only afterward do you "realize" why you blinked.

    a. What would you have to add to your diagram above to allow you to "realize" why you blinked?

    b. Why does it take you longer to "realize" why you blinked than it takes for the blink reaction to occur?

3. What composes the central nervous system (CNS), and how does the CNS differ in general form and function from the peripheral nervous system (PNS)?

4. The peripheral nervous system is subdivided into the somatic, autonomic, and enteric nervous systems.

    a. What is the general function of each?

PNS subdivision:	General function:
Somatic nervous system	
Autonomic nervous system	
Enteric nervous system	

    b. The autonomic system is again further subdivided into the sympathetic and parasympathetic systems. How do the general functions of each of these divisions interact to maintain homeostasis?

Autonomic nervous system divisions:	Functions:
Sympathetic	
Parasympathetic	

49.1 Test Your Understanding

Fibromyalgia affects about 2% of the U.S. population. Its primary symptoms are chronic pain, difficulty sleeping, and fatigue that is not relieved by sleep. Other reported symptoms include: jaw clenching at night, dryness of mouth and eyes, an increased need to urinate, irritable bowels, muscle numbness and tingling, and headaches.

1. Several researchers have proposed that fibromyalgia results from failed regulation of part of the autonomic nervous system. If this is true, failed regulation of which division of the autonomic nervous system could account for these symptoms? Explain your reasoning.

2. To test their ideas, the researchers recorded heart rate of both controls and individuals with fibromyalgia for 24-hour periods of normal activity. They observed that patients with fibromyalgia showed greatly elevated heart rates both while active and asleep. Does the addition of this information change or support the hypothesis you developed in 1, above?

Activity 50.1 How does sarcomere structure affect muscle function?

Working in groups of three or four, use playdough or cutout pieces of paper to construct a dynamic model of the sarcomere and demonstrate the sliding-filament model for muscle contraction.

When developing and explaining your model, be sure to include definitions or descriptions of the following terms and structures.

thick filament	sarcomere
thin filament	contracted length
actin	extended length
myosin	troponin complex
Z line	tropomyosin
H zone	Ca^{++}
cross-bridges	

Alternatively, assign at least three people to model different parts of the sarcomere and demonstrate its action in muscle contraction. Assign a different role to each person, and assign specific actions to each role.

When you have completed your model, explain it to another student or to your instructor.

Use your understanding of the model to answer the questions.

1. How are sarcomeres arranged within muscle fibers (and therefore within muscles)?

2. Describe the sliding-filament model of muscle action. For example, what interactions power the movement of filaments in association with each other? Include troponin complex, Ca^{++} binding sites, tropomyosin, myosin binding site, and actin in your discussion.

3. How do your answers to questions 1 and 2 help explain why, at maximum contraction, muscle measures 70% of its extended (uncontracted) length?

4. Review Figure 50.29 and the associated text in *Biology*, 8th edition. Then answer the next questions.

a. Acetylcholine, an excitatory neurotransmitter, is the neurotransmitter released into the neuromuscular junction. This release can trigger an action potential along the length of the muscle membrane. Describe the process involved.

b. If acetylcholine always produces an excitatory response in the neuromuscular junction, how can we regulate which muscles (in the arm, for example) are contracted and which are extended at any given time? For example, how can we bend the arm only partially?

c. Severe calcium deficiency can lead to a reduction in bone mass. It can also have serious effects on the functioning of the nervous system and on the action of muscles. Explain what role(s) calcium plays in the activity of muscles.

50.1 Test Your Understanding

The level of musculature in various organs is directly related to the amount of work done by that organ or organ part.

a. Give two examples of different parts of organ systems where this is evident. (Use two different organ systems.)	b. For each example, explain the advantage of the difference in musculature for each system.

Name_____   Course/Section_____

Date_____   Professor/TA_____

Activity 50.2 What would happen if you modified particular aspects of muscle function?

Test your understanding of muscle function by asking yourself what would happen if a specific part was damaged. What would the system still be able to do? What would it be unable to do?

In each of the following situations, indicate whether the proposed answers are true or false. If an answer is false, indicate why it is false.

1. A toxin in a newly discovered bacterial strain causes irreversible inactivation of the acetylcholine receptor. If you were infected with this organism, the symptoms would include:

T/F   a. Uncontrollable muscular contractions

   Explain:

T/F   b. Muscular paralysis

   Explain:

T/F   c. Loss of membrane potential of muscle fibers

   Explain:

2. In an experiment, a physiologist was able to destroy some but not all of the motor neurons to a specific skeletal muscle. Given what you know about motor units, what is the result of this action?

T/F   a. There is no effect because motor neurons are redundant; multiple neurons innervate each muscle fiber as backup.
       Explain:

T/F   b. The muscle will contract faster if the physiologist destroyed inhibitory motor neurons.
       Explain:

T/F   c. The muscle will, in general, contract more weakly because fewer fibers are capable of contracting.
       Explain:

T/F   d. The ability of the muscle to contract will not change because the electrical potential will spread from stimulated fibers to adjacent fibers.
       Explain:

Name_____ Course/Section_____

Date_____ Professor/TA_____

Activity 51.1 What determines behavior?

1. Some plant species (for example, many orchids) rely on a single species of insect for pollination. If the insect species dies out, so will the plant species. In a through c, refer only to the behavior of the insect species.

    a. What questions would you need to ask to determine proximate causation for this species-specific insect behavior?

    b. What questions would you need to ask to determine ultimate causation for this insect behavior?

    c. What kinds of experiment(s) or investigation(s) would you propose to answer at least one of the questions in parts a and b?

2. Many bird species that are common in the northern United States during spring and summer fly south in the fall to winter over and feed in Central or South America. In the spring, they return to the northern United States to breed.

   a. What questions would you need to ask to determine proximate causation for this behavior?

   b. What questions would you need to ask to determine ultimate causation for this behavior?

   c. What kinds of experiment(s) or investigation(s) would you propose to answer at least one of the questions in parts a and b?

3. Many species of animals engage in complex courtship rituals. Among these species is the bowerbird of Australia. Male bowerbirds construct elaborate structures, or bowers, from twigs, leaves, and moss and decorate them with colorful objects such as berries and shells. The bowers and the dances the males perform are designed to attract female bowerbirds for mating. After mating, the females fly away to build a nest and raise the offspring. The males remain at their bowers and try to attract additional mates. These birds can live for up to 17 years. Males are territorial and build their bowers in the same location each year. In studying their behavior, researchers have noted that about 25% of the females "shop around," going from one bower to another before deciding on a mate. The other 75% appear to go directly to a single bower to mate. These behaviors cannot be observed in captivity.

a. What questions would you need to ask to determine proximate causation for the bower-building behavior?

b. What questions would you need to ask to determine the proximate causation for the female choice?

c. What questions would you need to ask to determine ultimate causation for these behaviors?

d. What kinds of experiment(s) or investigation(s) would you propose to answer at least one of the questions in either part a or b?

Name_____   Course/Section_____

Date_____   Professor/TA_____

Activity 52.1 What factors determine climate?

The map on the next page shows a hypothetical continent on Earth. Assume that biomes and climates on this continent are produced by the same factors that produce biomes and climates on Earth's real continents. Use this map to answer the questions in this activity. Where needed, draw the required features directly on the map.

  1. a. On the map of the hypothetical continent, indicate the location(s) of the biomes listed in the table below. To do this, draw approximate boundary lines to delimit each biome type, and then label each delimited area with the type of biome it contains.

       b. In the table indicate the annual temperature and precipitation ranges for each biome.

Biome type	Annual temperature and precipitation ranges
a. Tropical rain forest(s)	
b. Temperate grasslands	
c. Chaparral	
d. Temperate forest(s)	
e. Tundra	

  2. Atmospheric circulation is driven primarily by differential heating of Earth's surface. More heat is delivered near the equator than near the poles. This seems to explain the northward and southward flows of air. What introduces the eastward and westward components into air movement? (*Hint:* Review Figure 52.10, Global Air Circulation and Precipitation Patterns and Global Wind Patterns, in *Biology,* 8th edition.)

  3. Use your understanding of global air circulation and wind patterns to draw arrows on the map of the hypothetical continent indicating:

       a. The direction of prevailing winds at points W_1, W_2, and W_3

       b. The direction of flow of surface currents in the ocean at points O_1, O_2, and O_3 (*Hint:* Note on the map in Figure 52.10 that surface currents in the ocean follow the major wind systems at the surface.)

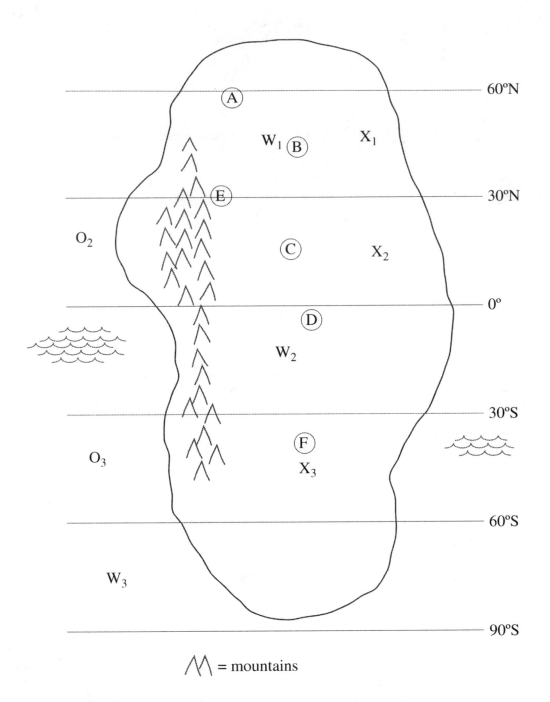

60°N

W₁ Ⓑ   X₁

Ⓐ

Ⓔ   30°N

O₂   Ⓒ   X₂

0°

Ⓓ

W₂

30°S

Ⓕ
X₃

O₃

60°S

W₃

90°S

$\bigwedge\bigwedge$ = mountains

A hypothetical continent

4. Are the surface winds at the given points warming or cooling as they move? Explain.

Point on the map	Are the winds warming or cooling as they move?	Explanation
a. X_1		
b. X_2		
c. X_3		

5. What biomes or vegetation types would most likely be found at the given points? (Assume all are at sea level or low altitudes.)

Point on the map	Most likely type of biome or vegetation
A	
B	
C	
D	

6. Would the climate at point E be relatively wet or dry? Explain.

7. What would the direction of the prevailing winds be at Earth's surface at point X_3?

8. In the United States, temperate forest extends from the East Coast westward for about 1,200 miles (to the Mississippi River Valley). From there, the forest begins to thin out toward the west into oak savannas (or temperate woodland), and it finally gives way to open grassland (Great Plains). The grasslands extend 1,000 miles westward to the foothills of the Rocky Mountains.

   a. Why does grassland replace forest west of the Mississippi River?

   b. What is the rain shadow effect?

   c. Draw a rain shadow somewhere in the *southern* hemisphere of the map.

9. What biome would exist at point F?

10. How are the general characteristics of plants (for example, morphology) influenced by climate? In other words, explain what effects climate has on the types of plants that grow in an area.

11. Refer to Figure 52.20 in *Biology*, 8th edition. In general, how is the distribution of major ecosystems or biomes related to climate? If you know the mean annual precipitation and the mean annual temperature of an area, would you be able to accurately predict the type of biome that could exist there? Explain.

12. Why isn't Earth's climate uniform? To answer this, summarize the major factors that can produce differences in climate from place to place.

Name_____   Course/Section_____

52.1 Test Your Understanding

1. A large asteroid hits the Earth, and alters its axis tilt from 23.5° to 10°.

   a. What effect(s) would this have on seasons in the northern hemisphere?

   b. What effect(s) would this have on biome distribution in the northern hemisphere? (Refer to Figure 52.20 in *Biology*, 8th edition.)

2. If you travel across the United States on a line from North Carolina to Southern California, you will find that temperate forest extends from the East Coast westward for approximately 1,200 miles to about the valley of the Mississippi River. There it begins to thin out toward the west into oak savannas and finally gives way entirely to open grassland (Great Plains), which extends another 1,000 miles westward up into the foothills of the Rocky Mountains. If you cross over the mountains you find chaparral in Southern California.

T/F The single most important factor affecting the type of biome present in each of these major regions of the United States is rainfall.

3 to 6. The type of vegetation that will grow in a particular region of the earth is strongly affected by:

T/F 3. Extremes of temperature (high and low) experienced over the year in that region.

T/F 4. Altitude of that region.

T/F 5. Seasonal availability of water.

T/F 6. Longitudinal location of that region.

Name_____   Course/Section_____

Date_____   Professor/TA_____

Activity 53.1 What methods can you use to determine population density and distribution?

1. To measure the population density of the chipmunks in a particular park, you sample several plots and capture 50 chipmunks. You mark each of their backs with a small dot of red paint and then release them. The next day, you capture another 50 chipmunks. Among the 50, you find 10 that are marked.

   a. Using the mark–recapture formula below, how many chipmunks do you estimate the population contains?

$$\frac{\text{Number of recaptures in second catch}}{\text{Total number in second catch}} = \frac{\text{Number marked in first catch}}{\text{Total population } N}$$

   b. What effect would each of the following discoveries have on your estimate?

      i. You later discover that you sampled the one area of the park that was most favored by the chipmunks.

      ii. You later discover that the chipmunks were licking the marks off each others' backs.

      iii. You later discover that the marked chipmunks are easier to see and therefore more susceptible to predation.

c. How could you modify your sampling program to ensure that you make more accurate estimates of population size?

2. Refer to the two proposals for the distribution of a tree species below.

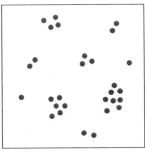

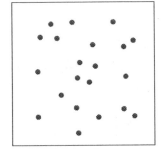

Proposed distribution 1   Proposed distribution 2

a. What type of distribution is represented in each of the proposals?

Distribution 1:

Distribution 2:

b. Given these two possible distributions, what factors do you need to consider in setting up a sampling plan for the area? Propose sampling strategies and the results you would get if organisms were distributed as in 1 vs. 2 above. For each sampling strategy proposed, indicate how will you know if you have chosen both an appropriate quadrat size and number of quadrats to provide you a good representation of both the size of the population and the actual distribution of organisms within the sampling area.

3. The following table shows the numbers of grasshopper deaths per acre per year resulting from two different agents of mortality applied to grasshopper populations of different densities.

a. Fill in the mortality rates for agents A and B in the table below.

Grasshopper population density (individuals/acre)	Deaths per year per acre		Mortality rate (%)	
	Agent A	**Agent B**	**Agent A**	**Agent B**
100	4	0		
1,000	40	25		
10,000	400	500		
100,000	4,000	50,000		

b. Graph the data below.

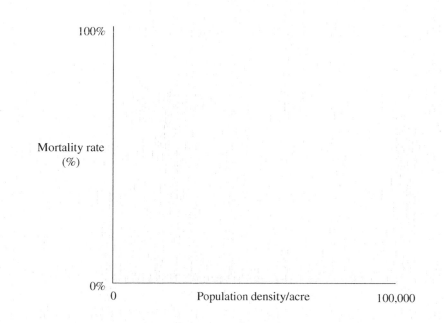

c. Which of the two agents of mortality (A or B) is operating in a density-independent manner? Explain your answer.

d. Which of the two agents of mortality (A or B) is likely to act as a factor stabilizing the size of the grasshopper population? Explain your answer.

A researcher has recently discovered three species of parasites (A, B, and C) that infect developing salamanders. He suspects that one or more of these species cause fatalities during salamander development and that the death rate varies with salamander population density. To test his hypothesis, the researcher sets up a series of experiments. In each experiment, he varies the density of the salamander populations. At the start of each experiment, he infects 5% of each test population with a single parasite species and then measures the mortality/death rate after four weeks. To establish a baseline mortality rate, he sets up a control experiment that differs only in that no parasite is introduced at any density. The data table and graph below relate the mortality rate of salamanders (caused by the three different parasites) to the original density of the developing salamander population.

a. Given the data presented in the graph, indicate whether each of the parasite species (A, B, and C) is acting in a density-dependent or density-independent fashion. Explain your answers.

Exp. 1		Exp. 2		Exp. 3		CONTROL	CONTROL
Density	A Mortality Rate	Density	B Mortality Rate	Density	C Mortality Rate	Density	Mortality Rate
20	0.1	20	0.82	20	0.4	20	0.1
40	0.2	40	0.74	40	0.25	40	0.096
80	0.4	80	0.76	80	0.49	80	0.12
160	0.6	160	0.8	160	0.35	160	0.15
320	0.85	320	0.75	320	0.36	320	0.092

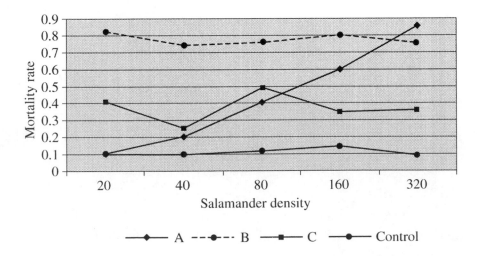

b. Looking at the data for parasites A and B, develop an argument to indicate which is more likely to cause extinction of the salamander population. Explain your reasoning.

Name_____   Course/Section_____

Date_____   Professor/TA_____

Activity 53.2   What models can you use to calculate how quickly a population can grow?

1. In the simplest population growth model, $dN/dt = rN$.

    a. What does each term stand for?

Term	Stands for
N	
dt	
r	
dN	

    b. What type of population growth does this equation describe?

    c. What assumptions are made to develop this equation?

2. Population growth may also be represented by the model $dN/dt = r_{max}N[(K - N)/K]$.

    a. What is K?

    b. If $N = K$, then what is dN/dt?

    c. Describe in words how dN/dt changes from when N is very small to when N is large relative to K.

    d. What assumptions are made to develop this equation?

3. You and your friends have monitored two populations of wild lupine for one entire reproductive cycle (June year 1 to June year 2). By carefully mapping, tagging, and censusing the plants throughout this period, you obtain the data listed in the chart.

Parameter	Population A	Population B
Initial number of plants	500	300
Number of new seedlings established	100	30
Number of the initial plants that died	20	100

a. Calculate the parameters for each population.

Parameter	Population A	Population B
B (births during time interval)		
D (deaths during time interval)		
b (per capita birth rate)		
m (per capita death rate)		
r (per capita rate of increase)		

b. Given the initial population size and assuming that the population is experiencing exponential growth at growth rate r, what will the number of plants be in each population in 5 years? (Use the initial population size as time 0 and compute to time 5.)

Population A:

Population B:

4. Using the exponential growth formula, you can determine the amount of time it will take for a population to double in size if you know r_{max} or r. The doubling time is equal to

$$\log_{10}2/\log_{10}(1 + r)$$

Alternatively, the doubling time per unit time can be estimated by using the formula:

70 divided by the percentage increase per unit time (as a whole number)

Using either of these formulas—the exponential growth formula or the approximate doubling rate formula—calculate the following.

a. If the population of a country is growing at 2% per year, how many years will it take for the population to double?

b. If your bank account is growing at the rate of 1% per year, how many years will it take for your money to double?

5. You are studying the growth of a particular strain of bacteria. You begin with a tiny colony on a petri plate. One day later, you determine that the colony grew and exactly doubled in size. A calculation showed that if the colony continued to grow at the same (constant) rate, it would cover the entire plate in 30 days. (Assume that colony size is directly proportional to the number of individual bacteria.)

   a. What is the value of r?

   b. On what day would the bacteria cover half the plate?

6. You collect data on birth and mortality in three populations of grasshoppers, and you calculate the following birth and death rates for these populations. Both populations are experiencing exponential growth:

	b	d
Population A	0.90	0.80
Population B	0.45	0.35
Population C	0.15	0.05

Are these statements true or false?

T/F   a. Population A is growing at the fastest rate.

T/F   b. Population C has the lowest death rate.

T/F   c. Population C is growing at the slowest rate.

T/F   d. All populations are growing at the same rate.

7. In a herd of bison, the number of calves born in 1992, 1993, and 1994 was 55, 80, and 70, respectively. In which year was the birth rate greatest?

8. A population of pigeons on the west side of town has a per capita annual growth rate of 0.07. A separate population of pigeons on the east side of town has a per capita annual growth rate of 0.10. If both populations are growing exponentially and both are censused the following year, in which of the populations will dN/dt be greatest?

9. Suppose you have a "farm" on which you grow, harvest, and sell edible freshwater fish. The growth of the fish population is logistic. You want to manage your harvest to maintain maximum yields (that is, the maximum rate of production) from your farm over a number of years.

As a fisheries manager, you are responsible for deciding how many walleye can be harvested without destabilizing the population.

a. Below is a data table showing the walleye population in a typical pond on your fish farm over 24 weeks. Draw a graph showing how population size in the pond changes through time.

Time (weeks)	1	2	3	4	5	6	7	8	9	10	11	12
Population size	100	101	102	103	104	106	110	115	125	140	155	172
Time (weeks)	13	14	15	16	17	18	19	20	21	22	23	24
Population size	188	201	209	217	221	225	229	233	235	237	238	239

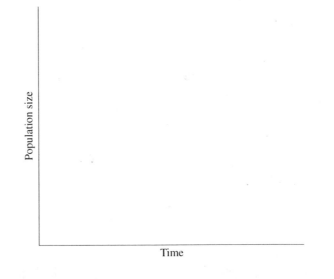

b. How large should you let the population get before you harvest? Identify the point on your graph *and explain why.*

c. Assume the carrying capacity for your pond is 250 individuals. Check your answer in part b by using the data in the chart and computing the change in the population size (dN/dt) when the population is at several different levels relative to its carrying capacity. Use $K = 250$ and $r_{max} = 0.20$.

Population size (N)	$(K - N)/K$	dN/dt
25 (low)		
50 (moderately low)		
125 (half K)		
200 (moderately high)		
250 (high)		

10. A rabbit population has the following life table.

Age class	Number of survivors	Number of deaths	Mortality rate	Number of offspring per reproducing pair
0–1	100	10	0.10	0
1–2	90		0.33	1.5
2–3	60	30		2.0
3–4	30	24	0.80	2.5
4–5		6	1.0	0

a. Fill in the missing data in the table.

b. Owing to a good food supply and a small predator population, the rabbit population is growing by leaps and bounds. The rabbits call a meeting to discuss population control measures. Two strategies are proposed:

- Delay all rabbit marriages until age class 2–3 (rabbits *never* breed until after marriage).
- Sterilize all rabbits in age class 3–4.

Which of the proposed strategies will be more effective in slowing population growth? Explain your reasoning and show your calculations.

Name_____   Course/Section_____

Date_____   Professor/TA_____

Activity 54.1 What do you need to consider when analyzing communities of organisms?

Understanding problems in community ecology most often requires the integration of a number of ecological principles.

For questions 1 to 5, analyze the situations described. Then explain which of the following ecological principles could be active in each particular situation.

Ecological principles:

coevolution   character displacement

realized niche   top–down vs. bottom–up controls

fundamental niche   trophic structure or trophic levels

tolerance   keystone species

competitive exclusion   competition

resource partitioning   disturbance

1. A small clan of hyenas killed an antelope. While they were feeding on the carcass, two female lions approached, growled at the hyenas, and chased them away from the carcass.

2. Two species of closely related swallows live in England. The black swallow lives in coniferous forests, and the yellow swallow lives in deciduous forests. In Ireland, where the black swallow has never been introduced, only the yellow swallow is present and it lives in both coniferous and deciduous forests.

3. In a woodland community, three species of rodents coexist: voles, field mice, and shrews. All three species eat seeds and nuts. Each species has a preference for seeds of the most appropriate sizes for their teeth and mouths; however, all three species compete for the same kinds of nuts. An owl species also lives in this woodland community. The owl preys on all three rodent species. During one particular year, a parasite that causes pneumonia in birds is introduced into the community. This parasite dramatically reduces the owl population, which remains low for several years as a result. Following the initial reduction in the owl population, there is a dramatic increase in the population of field mice and a dramatic decrease in the populations of both voles and shrews.

4. In 1962, five mute swans escaped from captivity and began a breeding population in Chesapeake Bay. Today, there are over 4,000 mute swans living in the bay. Each year they eat approximately 10.5 million pounds of aquatic grasses. These grasses provide habitat for waterfowl and crustaceans, improve water quality, decrease erosion, and increase dissolved oxygen concentrations in the bay. The swans are also aggressively territorial, and have been known to trample nests of other birds (e.g., least terns and black skimmers) and drive native birds such as tundra swans and black ducks from feeding and roosting areas.

5. Known as the "Hawaiian woodpecker," the `akiapola`au (aki-a-pul-a-ow) is found only in montane mesic old-growth koa/`ohi` forests, and only on the Big Island (Hawaii). It has a distinctive beak that is like a multiple-use tool.

The short straight lower mandible is used to peck holes in the wood and the long curved upper mandible is used to probe for insects and larvae. Males have larger beaks than females and feed on the trunks of trees. Females feed higher on branches and twigs. `Akiapola`au are thought to have the lowest reproduction rate for a small bird—only one chick per year, which is cared for by the parents for 6 months or more. The decline in their numbers appears to correspond with the introduction of rats, cats, and logging on the island.

6. A researcher collected data on an experiment she conducted on two desert islands. The islands were of similar size, climate, and species composition and richness, and were the same distance from the mainland. Originally, the same species of snake was present on both islands (A and B). In her experiment, the researcher removed the snake species from island A. For comparison, the snake species was NOT removed from island B. She then recorded the number of animal species on each island over a period of 24 months. Her data are presented in the table below.

a. Graph the data.

| Time (months) | # Species | |
	A	B
1	36	36
2	38	37
3	35	34
4	33	35
5	31	36
6	32	38
7	29	40
8	29	36
9	26	38
10	24	39
11	20	36
12	18	37
13	18	35
14	13	34
15	11	36
16	10	38
17	9	40
18	10	38
19	10	36
20	8	38
21	9	39
22	10	36
23	9	37
24	8	35

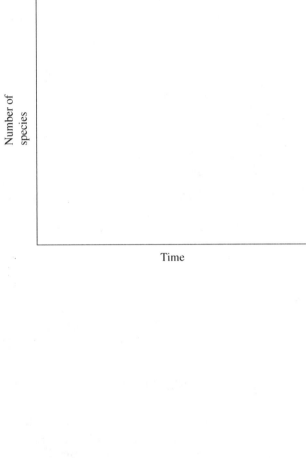

b. Construct a hypothesis that explains the difference between the numbers of species present on island A versus island B over the 24-month period.

c. Propose an experimental design to test your hypothesis. Explain the reasoning behind your design.

d. What would you expect to find as a result of your experiment? Describe your expected results and draw them on the graph below.

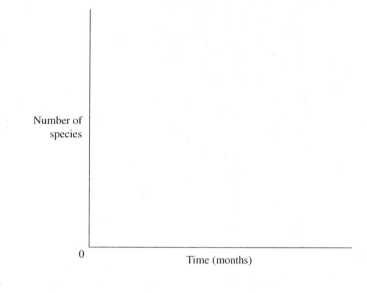

Name_____   Course/Section_____

Date_____   Professor/TA_____

Activity 54.2 What effects can a disturbance have on a community?

Disturbances rarely affect only one species in a community. More often, disturbances that have direct effects on a single species produce a cascade of additional effects on other species.

For questions 1 and 2, analyze the ecological situations described. Then answer the questions that follow each situation.

1. A particularly popular island vacation site is home to many species of orchids. The primary income-generating businesses on the island are tourism and orchid sales.

    Disturbance: To make the island more attractive to visitors, a local politician suggested that the island be periodically "fogged" with insecticides.

    Response: This fogging reduced the insect numbers. It also appeared to reduce the number of birds and new growths of orchids.

    (*Hint:* Examine the ideas of limits of tolerance, coevolution, predation, and bioaccumulation.)

    a. Is it likely that the disturbance directly caused the response?

    b. What other factors might be involved?

    c. How could you test different factors for their effect on the response? For example, what experiments could you set up?

d. What would you expect to find if the other factors you proposed affected the response?

2. Australia and New Zealand are home to a wide variety of marsupials (for example, kangaroos and other pouched mammals). Until colonization by foreign traders and other developments, placental mammals were not found in these areas.

Disturbance: Following colonization, the rabbit was introduced to Australia.

Response: The rabbit multiplied rapidly and ultimately became a pest species, doing considerable damage to both crop and natural plants.

(*Hint:* Examine the ideas of competition, competitive exclusion, predation, and coevolution.)

a. Is it likely that the disturbance directly caused the response?

b. What other factors might be involved?

c. How could you test different factors for their effect on the response? For example, what experiments could you set up?

d. What would you expect to find if the other factors you proposed affected the response?

3. In the boreal forest of Canada, wildfires are important disturbance factors. A single wildfire seldom burns a whole forest. Instead it burns large patches or stands and leaves others untouched. Following a wildfire in a black spruce forest, there is usually a predictable regrowth of the vegetation, starting with ground lichens and small spruce seedlings. As the spruce trees grow and form a closed-crown canopy, feather mosses (Bryophytes) are found in an increasing proportion on the forest floor. In some cases, the peat moss *Sphagnum* outcompetes the feather mosses and eventually dominates the ground cover. Because wildfires occur naturally about every 10 years, a forest stand can sometimes burn before *Sphagnum* dominance is reached, and the whole process repeats.

   a. Graph "Tree Biomass vs. Time" over a 100-year period:
      i.   at the stand scale (the stand is a particular part of the forest that burned) and
     ii.   at the landscape scale (composed of many forest stands).
        • Assume, at the stand scale, that when a stand burns, all trees in that stand die.
        • Assume, at the landscape scale, that there is one fire every 10 years, and that each fire burns a different stand.

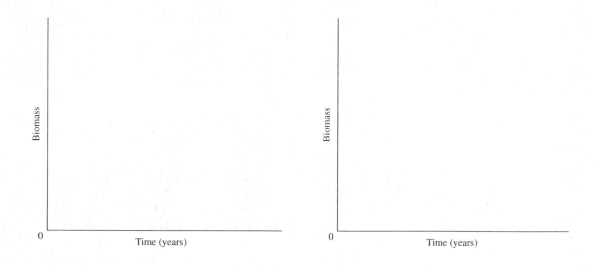

   b. Explain how (and why) the graphs differ.

Name_____   Course/Section_____

Date_____   Professor/TA_____

Activity 54.3 How can distance from the mainland and island size affect species richness?

1. Island biogeography theory attempts to explain the patterns of species richness and turnover on islands as a function of the size of the island and its distance from the mainland.

    a. What are the basic tenets of this theory? (Refer to Figure 54.27 in *Biology*, 8th edition.)

    b. In the graph below, curves A and A' represent the historic immigration and extinction rates, respectively, for an island off the coast of South America. Given these data, what is the equilibrium number of species for this island?

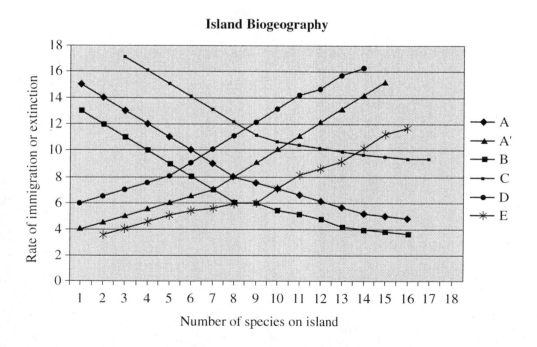

Island Biogeography

Legend: A, A', B, C, D, E

y-axis: Rate of immigration or extinction

x-axis: Number of species on island

c. If a land bridge or an isthmus forms and connects the island to the mainland, which two curves would best represent the resulting immigration and extinction rates for the island?

d. What would be the new equilibrium number of species?

2. A research scientist comes back from studying a group of islands in the South Pacific. He says his data show that, in some cases, the smaller islands had larger numbers of lizard species on them than did the larger islands. He suggests these data indicate that we should reexamine the whole theory of island biogeography.

a. What additional information would you like to have about the group of islands he studied?

b. Is there any way his data could be accounted for using the current theory of island biogeography? Explain your reasoning.

Name_____   Course/Section_____

Date_____   Professor/TA_____

Activity 55.1 What limits do available solar radiation and nutrients place on carrying capacities?

1. Energy transformations in a community can be diagrammatically represented as a trophic structure.

    a. Diagram a simple trophic structure for a grassland community.

    b. At what trophic level(s) do humans belong in your diagram? Explain.

2. Energy is lost at every step in the trophic structure where energy is transferred—for example, from primary producers to primary consumers. It is estimated that only 1% of the solar energy striking plants is converted to chemical energy, only 50–90% of gross primary productivity (GPP) becomes net primary productivity (NPP), and only about 10% of the energy available at each tropic level transfers to the next level (as biomass).

    a. Restate the preceding information in your own words. Be sure to explain what GPP and NPP are and how they relate to the percentage of energy transfers from one trophic level to another.

    b. Where does the "lost energy" go? Is it correct to say that energy is "lost" in these transfers? If not, what would be a better way of expressing this?

Copyright © Pearson Education, Inc., Publishing as Benjamin Cummings

3. *Biology*, 8th edition, indicates that each day Earth is "bombarded by about 10^{22} joules of solar radiation (1 J = 0.239 cal)." It has also been estimated that primary producers on Earth collectively create about 170 billion tons of organic material per year. Do the following calculations to determine how these two values compare.

   a. Convert 10^{22} joules per day to kilocalories per year.

   b. Convert 170 billion tons of organic material to kilocalories. To do this, assume all of the 170 billion tons is glucose. Also assume tons are metric tons. (*Note:* 180 grams of glucose burned in a calorimeter gives off 686 kcal of energy.)

   c. Given your answers to parts a and b, what percentage of the total incoming solar energy is captured as biomass (glucose)?

   d. Measurements of photosynthetic conversion of the energy in sunlight to biomass indicate that, at best, plants can convert about 75% of the energy they absorb as sunlight into biomass. The other 25% is used to support metabolism. What could account for the apparently low efficiency of photosynthesis relative to the total incoming solar energy that you calculated in part c?

4. If we can calculate Earth's total primary productivity, we can use it to develop an estimate of the total number of humans Earth can support.

    a. How much energy in kilocalories per year would it take to support Earth's current human population if all individuals weighed about 75 kg and each required 2,000 kcal per day? Assume 6×10^9 humans currently inhabit Earth.

    b. How does the estimate of the energy contained in the 170 billion tons of organic material (from part b in question 3) compare to the amount of energy required to support the current human population (from part a in question 4)?

    c. Is it reasonable to assume that all of the primary productivity on Earth is available to support humans? If not, what else do you need to consider?

5. You have been monitoring the net primary productivity (NPP) of a grassland area for several years. Over the years, NPP increased initially and then leveled off. You suspect that the availability of a nutrient is limiting productivity.

a. Design an experiment to determine whether there is a limiting nutrient.

b. Given your design, what results would you expect if only one of the nutrients you test is limiting?

c. What results would you expect if more than one of the nutrients are limiting? Would this affect your design in any way?

d. What factor(s) other than nutrient limitation might cause NPP to level off?

55.1 Test Your Understanding

1 to 3. Two species of hawks are found at various times of the year in the United States. Hawk species A eats mice and squirrels (both of which eat seeds). Hawk species B eats snakes that feed on mice and squirrels. Given what you know about trophic structures, if other characteristics of the two hawk species are similar, which of the following are likely to be true? Explain your reasoning.

A = Likely to be true.
B = Not likely to be true.

1. For a given feeding area, the number of hawk species A that could be supported is likely to be smaller than the number of hawk species B that could be supported.

2. Reduction in the available primary productivity in the feeding area would tend to have a lesser effect on hawk species A than on hawk species B.

3. If the seeds consumed by the mice and squirrels in this feeding area were contaminated with heavy metals, you would expect to find much lower concentrations of heavy metals in hawk species A than in hawk species B.

Name_____   Course/Section_____

Date_____   Professor/TA_____

Activity 56.1 What factors can affect the survival of a species or community?

Making decisions to preserve communities requires an understanding and integration of many factors. Assume you work for the U.S. government and you manage a large national forest. You are told that to maintain the economy in the area, the government has agreed to allow foresters to remove half a million acres of trees from a million-acre parcel. This parcel is almost square.

You have asked your staff to do an analysis of two possible methods for implementing the plan:

- Proposal I: Split the million acres into two parcels of a half million acres each and allow the foresters to harvest all trees on one of these parcels.
- Proposal II: Divide the million-acre tract into 50 parcels of 10,000 acres each and allow the foresters to cut half the trees in each parcel.

1. List some of the ecological advantages and disadvantages of each proposal.

Proposal I		Proposal II	
Advantages	Disadvantages	Advantages	Disadvantages

2. Given the characteristics of the various animal species described below, which of the forest-cutting proposals would be more likely to ensure the continued success of each animal species? Explain your answers.

a. *C. arnivora* is a secondary and occasionally a tertiary consumer or carnivore. Behaviorally, it roams over about 20 square miles of "home range" in search of food—for example, rabbits (herbivores) and foxes (carnivores).

b. *R. odentia* is a small rodentlike herbivore that exists in small numbers in the forest. Its preferred food and habitat are found along the edges of the forest and are composed primarily of herbaceous (nonwoody) annual plants that produce tender shoots in the spring and plentiful seeds later in the year.

c. *P. redatoria* is a predatory bird that feeds on small rodents and occasionally on snakes and other reptiles. It can range over large distances looking for food. It nests in hollows that form naturally in a particular species of tree. These hollows are not found in trees under 20 years of age but are common in trees 40 years and older.

3. Conservation biologists have debated extensively which is better: many small reserves or a few large ones.

a. What factors should be considered in making judgments about the size and location of reserves?

b. Some ecologists argue that we should be concerned about preserving the largest number of species. Others argue that we should be most concerned with saving those species judged to be of unusual importance. Develop an argument to support one of these viewpoints. Your argument should be based on our existing understanding of biology in general and ecology in particular.

4. Based on your answers to questions 1–3, which of the proposals for removing half a million acres from the national forest in question 1 would you recommend? Explain your reasoning.

Appendix A   An Introduction to Data Analysis and Graphing for a PC

This appendix provides a quick introduction to the use of some of the data analysis and graphing capabilities of one of the more commonly used computer programs, Excel. This and other similar software systems make it much easier to graph and analyze large data sets.

1. TO SET UP EXCEL FOR DATA ANALYSIS AND GRAPHING

- Open Excel and activate the "Data Analysis" tool.
    - Under the "Tools" pull-down menu, look for "Data Analysis."
    - If you don't see "Data Analysis," look for "Add-Ins." Click on "Data Analysis" or "Add-Ins" and select "Analysis ToolPack"; then click "OK."
    - If you did a custom install instead of a complete install of Excel, you may not find "Analysis ToolPack" under "Add-Ins." If this is the case, go to your original installation CD to add this feature.
- Enter the data to be analyzed or graphed in columns on the Excel spreadsheet.
    - If you want to graph height versus arm length, enter the data in two columns with the data you want to appear on the y-axis (independent variable) of a graph in the right-hand column.
    - The data to appear on the x-axis (dependent variable) should be in the adjacent left-hand column.
    - An example is on the next page.

Note: If you are using Excel 2007 to find and activate Add-ins:

- Click on the "Office" button in the upper left-hand corner.
- Then click on the "Excel options" button at the bottom of the pop-up window.
- Select "Add-ins" from the left-hand column.
- Select/highlight "Analysis ToolPak" in the list of Inactive Application Add-ins, then click "Go."
- Check the box in front of "Analysis Tool Pak" in the window that appears and click "OK."
- Now click on the "Data" tab at the top of the screen. You should see a new "Data Analysis" pull-down menu at the right top of the screen.

	A	B	C
1	Gender	Arm Length (cm)	Height (cm)
2			
3	f	73.7	170.2
4	m	76.7	177.8
5	m	71.1	170.2
6	f	66	165.1
7	f	68.6	172.7
8	m	76.7	172.7
9	m	78.7	177.8
10	m	81.3	182.9

2. TO SORT THE DATA

- In the upper left-hand corner of the Excel spreadsheet, click the box between Row 1 and Column A. The whole table should be highlighted when you do this.
- Pull down the "Data" menu at the top of the screen and click "Sort."
- If you have headers or titles on your columns, click the ○ in front of "Header row."
- Select the column you want to sort by (e.g., Gender).
- Select either "Ascending" if you want the sort to go from low to high (or A to Z) or "Descending" for high to low (or Z to A).
- Click "OK."

3. TO GET DESCRIPTIVE/SUMMARY STATISTICS FOR THE DATA

a. Pull down the "Tools" menu at the top of the Excel spreadsheet and click on "Data Analysis."

b. Click on "Descriptive Statistics" and then "OK."

- In the window that appears, for input range click on the icon to the right of the "Input Range:" box.
- Highlight (select) the data in the Excel spreadsheet that you want analyzed. For example, select all of the arm length data (males plus females).
- Click on the chart icon next to "Input Range" again to return to the Descriptive Statistics window.
- To indicate where to place the output, click on the ○ in front of "Output Range:" and then click on the chart icon to the right of the "Output Range:" box.
  - To select the location for the data to appear, click on any open box in the Excel spreadsheet.
  - Click the chart icon next to "Output Range" again to return to the Descriptive Statistics window.
- In the Descriptive Statistics window select:
  - ☐ Summary Statistics
  - ☐ Confidence Level for Mean
- Click "OK." A box containing the summary statistics will appear. For example:

Arm Length			Definitions:
Mean		74.1	The average of all the values.
Standard Error		1.849421	An estimate of the standard deviation of the sample mean.
Median		75.2	The number that lies in the middle of the set of numbers.
Mode		76.7	The most frequent number in the set.
Standard Deviation		5.230952	The square root of the variance.
Sample Variance		27.36286	A measure of how far from the average the data points are.
Kurtosis		–0.99518	A measure of how peaked or flat the distribution is.
Skewness		–0.29163	A measure of the degree of symmetry of the distribution.
Range		15.3	The difference between the lowest and highest values in the set.
Minimum		66	The smallest value in the set.
Maximum		81.3	The largest value in the set.
Sum		592.8	The total of all values in the set.
Count		8	The total number of values in the set.

Two data sets may have the same mean but very different amounts of variability. To indicate the amount of variability that exists in a given set of data, the arithmetic mean for those data is often expressed as the mean ± the standard deviation. Here, the smaller the standard deviation is, the smaller the variance is in the treatment group. The smaller the variance is, the more confidence we have that the data in our sample set represent a good estimate of the values for the larger population.

4. TO GRAPH THE DATA

a. To make a linear graph of x vs. y values:
- Click on the "Chart Wizard" icon at the top of the Excel spreadsheet.
- Click on "XY (Scatter)."
- Click "Next."
- Click the chart icon to the left of "Data range:"
  - Select all of the data in the two columns (Arm length [cm] and Height [cm]).

- - *Note:* If your data are not in side-by-side columns with the *y* data in the right-hand column, you can click on the "Series" tab instead of the "Data range:" tab. Click "Add" and then enter the data separately for the *x*- and *y*-axes by highlighting the columns separately here.
- Click "Next" and select the "Titles" tab to add titles for the chart and the *x*- and *y*-axes.
- Click "Next" and select where you want the chart to appear (on the same sheet or a new sheet). Then click "Finish."
- On the graph that appears, click on one of the data points to highlight it, then right click (on PC computers) and select "Add trendline."
  - Under the "Type" tab, select "Linear."
  - Under the "Options" tab, select:
    - "Display equation on chart" and "Display R-squared value on chart."
    - Click "OK."
- For the data in the previous table you should get the following graph.

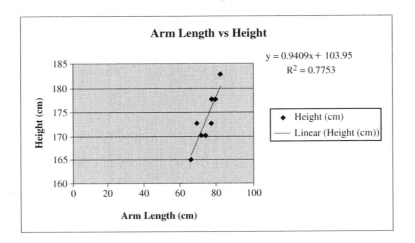

The formula $y = 0.9409x + 103.95$ is in the format $y = mx + b$. When the value of *m* is positive it indicates that there is a direct relationship between *x* and *y*; in this case it indicates that as *x* increases, *y* increases. (If *m* were negative, it would indicate an inverse relationship—as *x* increases, *y* decreases.)

For this specific graph, if it were reasonable to extrapolate the line to a point where $x = 0$, the line would intercept the *y* axis at *b*, or 103.95 cm. With this as the zero point for *x* on the line, the *y* values would increase from this point by *m*, or 0.9409 for every unit change in *x*.

The R^2 value has no units and a range of 0 to 1. An R^2 value of zero means there is no linear relationship between *x* and *y* values. On the other hand, high values of R^2 indicate a strong correlation between the *x* and *y* values. High R^2 values are also a good indicator of statistical significance.

b. To make a histogram of the data:
   • Select a set of data to use (e.g., heights for males only). (See previous section "To Sort the Data.")
   • Make a new column of data containing heights increasing by 5 cm each—for example, 160, 165, 170, 175, 180, 185 cm. (This is called your Bin Range.)
   • Go to the "Tools" pull-down menu and click on "Data Analysis," then on "Histogram." Then click "OK."
       ○ Click on the chart icon to the right of the "Input Range:" box.
           □ Select/highlight the height data for males only (include a column heading).
           □ Click the chart icon again to return to the Histogram window.
       ○ Click on the chart icon to the right of the "Bin Range:" box.
           □ Select/highlight the column containing heights increasing by 5 cm each.
           □ Click on the chart icon again to return to the Histogram window.
       ○ Click on the ○ in front of "Output Range:" and then click on the chart icon to the right of the "Output Range:" box.
           □ To select the location for the data to appear, click on any open box in the Excel spreadsheet.
           □ Click the chart icon again to return to the Histogram window.
       ○ In the Histogram window select:
           □ Chart Output
       ○ Click "OK." The frequency data and histogram should appear.

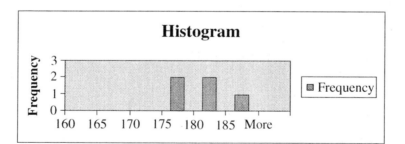

Histograms are often developed to help determine whether or not the data appear to be distributed normally. In this case, only five values (the five heights for males) are included—enough to determine how the heights among these five individuals are distributed but not enough to make any general statements about the total population of males.

Note: If the graph of the data approximates a bell-shaped curve or normal distribution, the standard deviation takes on additional meaning. For normal data distributions, about 68% of the data points will fall within one standard deviation (on either side) of the mean. Ninety-five percent of the data will fall within two standard deviations of the mean; and 99% will fall within three standard deviations of the mean.

5. TO DETERMINE STATISTICAL SIGNIFICANCE

The statistical significance of the difference in means can be determined by using a statistical test. The null hypothesis (H_0) is generally assumed for such tests. A simple null hypothesis would state that there is no difference in x (some characteristic) between two populations (e.g., there is no difference in arm length between males and females). This null hypothesis further assumes that any differences observed between the two populations are the result of chance variation or sampling error alone. An alternative hypothesis (H_a) would be that there **is** a difference or relationship between the populations. The alternative hypothesis is supported when the null hypothesis is rejected.

The Data Analysis function of Excel allows you to perform a number of different statistical tests, including t tests and analysis of variance (ANOVA). These tests are designed to show whether the observed difference in means, between groups or between control and test groups, is due to random chance/variability. A significance level of 0.05 or 5% means that statistically there is only a 5% chance that the difference observed is due to chance alone. A significance level of 0.05 is usually accepted as evidence that the treatment produced a significant effect. A significance level of 0.01, or 1%, is considered highly significant. It would mean that only 1% of the time would such a difference be likely to occur as a result of chance alone. In these cases, the null hypothesis is rejected.

Appendix B An Introduction to Data Analysis and Graphing for a Mac

This appendix provides a quick introduction to the use of some of the data analysis and graphing capabilities of one of the more commonly used computer programs, Excel. This and other similar software systems make it much easier to graph and analyze large data sets.

1. TO SET UP EXCEL FOR DATA ANALYSIS AND GRAPHING
- Open Excel and activate the "Data Analysis" tool.
  - Under the "Tools" pull-down menu, look for "Data Analysis."
  - If you don't see "Data Analysis," look for "Add-Ins." Click on "Data Analysis" or "Add-Ins" and select "Analysis ToolPack"; then click "OK."
  - If you did a custom install instead of a complete install of Excel, you may not find "Analysis ToolPack" under "Add-Ins." If this is the case, go to your original installation CD to add this feature.
- Enter the data to be analyzed or graphed in columns on the Excel spreadsheet.
  - If you want to graph height versus arm length, enter the data in two columns with the data you want to appear on the y-axis (independent variable) of a graph in the right-hand column.
  - The data to appear on the x-axis (dependent variable) should be in the adjacent left-hand column.
  - For example:

	A	B	C
	Gender	Arm Length (cm)	Height (cm)
1			
2			
3	f	73.7	170.2
4	m	76.7	177.8
5	m	71.1	170.2
6	f	66	165.1
7	f	68.6	172.7
8	m	76.7	172.7
9	m	78.7	177.8
10	m	81.3	182.9

2. TO SORT THE DATA

- In the upper left-hand corner of the Excel spreadsheet, click the box between Row 1 and Column A. The whole table should be highlighted when you do this.
- Pull down the "Data" menu at the top of the screen and click "Sort."
- If you have headers or titles on your columns, click the ○ in front of "Header row."
- Select the column you want to sort by (e.g., Gender).
- Select either "Ascending" if you want the sort to go from low to high (or A to Z) or "Descending" for high to low (or Z to A).
- Click "OK."

3. TO GET DESCRIPTIVE/SUMMARY STATISTICS FOR THE DATA

a. Pull down the "Tools" menu at the top of the Excel spreadsheet and click on "Data Analysis."

b. Click on "Descriptive Statistics" and then "OK."

- In the window that appears, for input range click on the icon to the right of the "Input Range:" box.
- Highlight (select) the data in the Excel spreadsheet that you want analyzed. For example, select all of the arm length data (males plus females).
- Click on the icon to the right of "Input Range" again to return to the Descriptive Statistics window.
- To indicate where to place the output, click on the ○ in front of "Output Range:" and then click on the icon to the right of the "Output Range:" box.
  - ○ To select the location for the data to appear, click on any open box in the Excel spreadsheet.
  - ○ Click the icon next to "Output Range" again to return to the Descriptive Statistics window.
- In the Descriptive Statistics window select:
  - □ Summary Statistics
  - □ Confidence Level for Mean
- Click "OK." A box containing the summary statistics will appear. For example:

Arm Length		Definitions:
Mean	74.1	The average of all the values.
Standard Error	1.849421	An estimate of the standard deviation of the sample mean.
Median	75.2	The number that lies in the middle of the set of numbers.
Mode	76.7	The most frequent number in the set.
Standard Deviation	5.230952	The square root of the variance.
Sample Variance	27.36286	A measure of how far from the average the data points are.
Kurtosis	–0.99518	A measure of how peaked or flat the distribution is.
Skewness	–0.29163	A measure of the degree of symmetry of the distribution.
Range	15.3	The difference between the lowest and highest values in the set.
Minimum	66	The smallest value in the set.
Maximum	81.3	The largest value in the set.
Sum	592.8	The total of all values in the set.
Count	8	The total number of values in the set.

Two data sets may have the same mean but very different amounts of variability. To indicate the amount of variability that exists in a given set of data, the arithmetic mean for those data is often expressed as the mean ± the standard deviation. Here, the smaller the standard deviation is, the smaller the variance is in our set of results or data. The smaller the variance is, the more confidence we have that the data in our sample set represent a good estimate of the values for the larger population.

4. TO GRAPH THE DATA
a. To make a linear graph of *x* vs. *y* values:
- Click on the "Chart Wizard" icon at the top of the Excel spreadsheet.
- Click on "XY (Scatter)."
- Click "Next."
- Click the icon to the right of "Data range:"
  - Select all of the data in the two columns (Arm length [cm] and Height [cm]).
  - *Note:* If your data are not in side-by-side columns with the *y* data in the right-hand column, you can click on the "Series" tab at the top of the screen

that pops up (instead of the "Data range" tab). Click "Add" and then enter the data separately for the x- and y-axes by highlighting the columns separately here.

- Click "Next" and select the "Titles" tab to add titles for the chart and the x- and y-axes.
- Click "Next" and select where you want the chart to appear (on the same sheet or a new sheet). Then click "Finish."
- On the graph that appears, click on one of the data points to highlight it, then click on the "Chart" pull-down menu at the top of the screen and select "Add trendline."
  - Under the "Type" tab, select "Linear."
  - Under the "Options" tab, select:
    - "Display equation on chart" and "Display R-squared value on chart."
    - Click "OK."
- For the data in the previous table you should get the following graph.

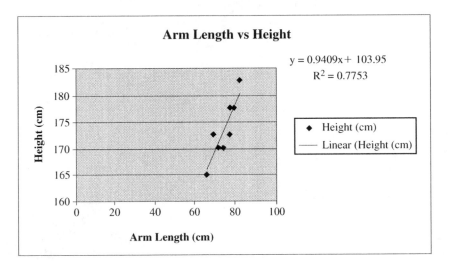

The formula $y = 0.9409x + 103.95$ is in the format $y = mx + b$. When the value of m is positive it indicates that there is a direct relationship between x and y; in this case it indicates that as x increases, y increases. (If m were negative, it would indicate an inverse relationship—as x increases, y decreases.)

For this specific graph, if it were reasonable to extrapolate the line to a point where $x = 0$, the line would intercept the y-axis at b, or 103.95 cm. With this as the zero point for x on the line, the y values would increase from this point by m, or 0.9409 for every unit change in x.

The R^2 value has no units and a range of 0 to 1. An R^2 value of zero means there is no linear relationship between x and y values. On the other hand, high values of R^2 indicate a strong correlation between the x and y values. High R^2 values are also a good indicator of statistical significance.

b. To make a histogram of the data:
- Select a set of data to use (e.g., heights for males only). (See the previous section "To Sort the Data.")
- Make a new column of data containing heights increasing by 5 cm each for example, 160, 165, 170, 175, 180, 185 cm. (This is called your Bin Range.)
- Go to the "Tools" pull-down menu and click on "Data Analysis," then on "Histogram." Then click "OK."
  - Click on the icon to the right of the "Input Range:" box
    - Select/highlight the height data for males only (include a column heading).
    - Click the chart icon again to return to the Histogram window.
  - Click on the icon to the right of the "Bin Range:" box.
    - Select/highlight the column containing heights increasing by 5 cm each.
    - Click on the icon again to return to the Histogram window.
  - Click on the ○ in front of "Output Range:" and then click on the icon to the right of the "Output Range:" box.
    - To select the location for the data to appear, click on any open box in the Excel spreadsheet.
    - Click the icon again to return to the Histogram window.
  - In the Histogram window select:
    - ☐ Chart Output
  - Click "OK." The frequency data and histogram should appear.

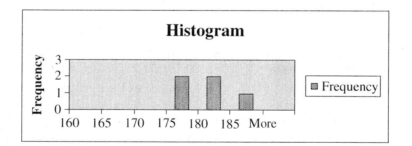

Histograms are often developed to help determine whether or not the data appear to be distributed normally. In this case, only five values (the five heights for males) are included—enough to determine how the heights among these five individuals are distributed but not enough to make any general statements about the total population of males.

Note: If the graph of a large data set approximates a bell-shaped curve or normal distribution, the standard deviation takes on additional meaning. For normal data distributions, about 68% of the data points will fall within one standard deviation (on either side) of the mean. Ninety-five percent of the data will fall within two standard deviations of the mean; and 99% will fall within three standard deviations of the mean.

5. TO DETERMINE STATISTICAL SIGNIFICANCE

The statistical significance of the difference in means can be determined by using a statistical test. The null hypothesis (H_0) is generally assumed for such tests. A simple null hypothesis would state that there is no difference in x (some characteristic) between two populations (e.g., there is no difference in arm length between males and females). This null hypothesis further assumes that any differences observed between the two populations are the result of chance variation or sampling error alone. An alternative hypothesis (H_a) would be that there **is** a difference in arm length (or other characteristic) between the populations. The alternative hypothesis is supported when the null hypothesis is rejected.

The Data Analysis function of Excel allows you to perform a number of different statistical tests, including t tests and analysis of variance (ANOVA). These tests are designed to show whether the observed difference in means, between groups or between control and test groups, is due to random chance/variability. A significance level of 0.05 or 5% means that statistically there is only a 5% chance that the difference observed is due to chance alone. A significance level of 0.05 is usually accepted as evidence that the treatment produced a significant effect. A significance level of 0.01, or 1%, is considered highly significant. It would mean that only 1% of the time would such a difference be likely to occur as a result of chance alone. In these cases, the null hypothesis is rejected.